ARTS

A SCIENCE MATTER

Science Matters Series

Lui Lam
Founder and Editor

Science Matters (SciMat) is the new discipline that treats all human-dependent matters as part of science, wherein, humans (the material system of *Homo sapiens*) are studied scientifically from the perspective of complex systems. That "everything in Nature is part of science" was well recognized by Aristotle and da Vinci and many others. Yet, it is only recently, with the advent of modern science and experiences gathered in the study of evolutionary and cognitive sciences, neuroscience, statistical physics, complex systems and other disciplines, that we know how the human-related disciplines can be studied scientifically. Science Matters Series covers new developments in all the topics in humanities and social sciences from the SciMat perspective, with emphasis on the humanities.

Published

1. *Science Matters: Humanities as Complex Systems*
   M. Burguete & L. Lam, editors

2. *Arts: A Science Matter*
   M. Burguete & L. Lam, editors

Science Matters Series

ARTS

A SCIENCE MATTER

Maria Burguete

Scientific Research Institute Bento da Rocha Cabral, Portugal

Lui Lam

San Jose State University, USA

Editors

World Scientific

NEW JERSEY • LONDON • SINGAPORE • BEIJING • SHANGHAI • HONG KONG • TAIPEI • CHENNAI

Published by

World Scientific Publishing Co. Pte. Ltd.
5 Toh Tuck Link, Singapore 596224
USA office: 27 Warren Street, Suite 401-402, Hackensack, NJ 07601
UK office: 57 Shelton Street, Covent Garden, London WC2H 9HE

British Library Cataloguing-in-Publication Data
A catalogue record for this book is available from the British Library.

ARTS: A SCIENCE MATTER

Cover design: Lui Lam

Artwork: Cover: Studying a Falling Apple (Lui Lam, 2010)
Part I: Gingko (Charlene Lam, 2010)
Part II: Birch (Charlene Lam, 2010)
Part III: Lilac (Charlene Lam, 2010)

ISBN-13 978-981-4324-93-9
ISBN-10 981-4324-93-0

Printed in Singapore by World Scientific Printers.

Preface

Science Matters is the new discipline that treats all human-dependent matters as part of science, wherein, humans (the material system of *Homo sapiens*) are studied scientifically from the perspective of complex systems. That "everything in Nature is part of science" was well recognized by Aristotle and da Vinci and many others. Yet, it is only recently, with the advent of modern science and experiences gathered in the study of evolutionary and cognitive sciences, neuroscience, statistical physics, complex systems and other disciplines, that we know how the human-related disciplines can be studied scientifically.

Science Matters (SciMat) covers all the topics in humanities and social sciences, arts in particular. Arts here include visual arts, literature, film, music, architecture, performance arts, new media arts and so on.

This book treats arts as part of science, from the unified perspective of SciMat. It is probably the first and only book to which academic professionals and practicing artists contribute, as equals, on the common theme of creating and understanding arts. It contains 17 chapters, with 18 contributors who are prominent humanists, professional artists or scientists. It consists of three parts: Part I: Philosophy and History of Arts; Part II: Arts in Action; Part III: Understanding Arts. The book is aimed at both research scholars and laypeople. While the discussions presented in the chapters of this book are very general and definitely applicable to all kinds of arts, for practical reasons, specific examples are mostly confined to visual arts, literature and film. We hope to cover other parts of arts in the future.

Two other features of this book should be mentioned. First, arts studies, like in any other discipline, can be and are carried out with three different approaches: empirical, phenomenological and bottom-up. (For the humanities, with arts as a particular case, the bottom-up approach could be starting from the neuro or the genetic level.) All these three approached in arts studies are represented in this book. Second, the general nature and the origin of arts, an unsettled problem for 2,400 years since Plato's time, are addressed in four chapters in this book. In particular, a plausible answer to this important problem is presented for the first time (Chapter 1).

It is our wish that this book will help to start a new trend in arts studies; that is, arts scholars and practicing artists work together, treating arts as part of science.

Rio Maior, Portugal — Maria Burguete
San Jose, California — Lui Lam

Contents Summary

Contents

PART I PHILOSOPHY AND HISTORY OF ARTS

1

Arts: A Science Matter

Lui Lam

The nature and origin of arts, and its relationship to "science" have been under much debate since Plato about 2,400 years ago. Here, a new perspective on these issues is presented. Science is to understand how nature works, while nature consists of (human and nonhuman) living systems and nonliving systems. Consequently all human-dependent matters are part of science—the premise underlying the new discipline called *Science Matters* (SciMat), which covers all topics in humanities and social sciences, arts in particular. (Arts here refer to visual arts, literature, film, performance arts, music, architecture, new media arts and so on.) In fact, arts are a subset of humans' creative activities that aim to excite the receiver's neurons in a certain manner, through that person's senses, with or without significant consequences. The usual kind of "science" is to understand mostly inanimate, simple systems and how the world/universe works; it is part of science in general. Arts as a science matter is to find out everything about arts, including arts' origin and nature, and how and why arts work at both ends of the creator and the receiver. Like physics and any other discipline, arts can be classified into two types—*pure arts* and *applied arts*. Some arts, such as drawing and performance art, could start a million years ago. All arts evolved over time and space, and the contents kept on changing as humans invented language and writing and as they migrated out of Africa and spread over the world; arts contain both global universal elements and local features. Here, all these issues as well as how arts as a science matter could be studied are elaborated, after a brief introduction to SciMat and humans' development history and inheritance mechanism (genes and epigenes) is given.

1.1 Introduction

Arts in this chapter refer to visual arts, literature, film, performance arts, music, architecture, new media arts and so on. The origin and nature of arts, and its relationship to "science"[1] are under much debate.[2] The confusion arises from many factors which will become clear later as our discussion goes along.

Here we will try to clear up these confusions by reexamining the problem and presenting a new perspective on all these issues. We will first clear up the definition of the word science and introduce the new discipline *Science Matters* [Lam, 2008a] within which arts belong (Section 2). Since arts are human activities, it is important to understand where we came from and how we developed evolutionarily (Section 3). The origin and nature of arts are then discussed, respectively, in Sections 4 and 5. Arts as a science matter are presented in Section 6, where the relationship of arts to "science" and how arts could be studied are given. Discussion and conclusion in Section 7 conclude the chapter.

1.2 Science and Science Matters

The scope and nature of science, and the new discipline Science Matters are presented here.

1.2.1 *What Is Science?*

Science is about the study of nature and a means to understand it in a unified way. Nature consists of everything in the universe—all material systems: humans and nonhumans. Consequently, the only logical conclusion about the scope of science is:

$$\begin{aligned} \text{science} &= \text{natural science} \\ &= \text{“physical” science} + \text{social science} + \text{humanities} \end{aligned} \tag{1.1}$$

---

[1] In this chapter, "science" with quotation marks means the science of mostly inanimate systems (i.e., science in the narrow sense), identical to the conventional usage of the word.

[2] There is a sizable body of literature on these topics. At the entry level, see [Appenzeller, 1998; Brown & Dissanayake, 2009] for arts' origin, [Carroll, 1999; Adams, 1996; Dutton, 2009] for arts' nature, and [Strosberg, 2001; Miller, 1996; Leibowitz, 2008] for arts and "science."

where the three items on the right-hand side of Eq. (1.1) are in decreasing level of scientific development; they are *not* classified according to the nature of the objects under study [Lam, 2008a].

That "everything in nature is part of science" was well recognized by Aristotle and da Vinci and many others. However, it is only recently, with the advent of modern science and experiences gathered in the study of evolutionary and cognitive sciences, statistical physics [Lam, 1998; Paul & Baschnagel, 1999], complex systems [Lam, 1997; 1998] and other disciplines, that we know how the human-related disciplines can be studied scientifically.

1.2.2 *Three Misconceptions about Science*

The contents of science can be divided into two parts: human-dependent and human-independent. The study of the human-dependent part (humanities and social science) was hindered by three misconceptions in science. In fact, the miserable part of human history (e.g., ideological massacre and religious burning at stake) is related to these three misconceptions.

Misconception 1: Natural systems include non-human systems only (i.e., humans are excluded). This misconception started, at least, from the early Greek time some 2,400 years ago. It is wrong because all material systems in nature, humans included, are made up of atoms created in the stars some 300 million years ago [Turner, 2009].

Misconception 2: Physics is about deterministic systems only (i.e., stochastic systems[3] are excluded). This misconception is due to the tremendous success of Newtonian physics in the past 300 years and the ignorance of physics developments (Fig. 1.1).

---

[3]Stochastic is a technical word in physics, meaning that probability appears somewhere in the process; a random process is a special case [Paul & Baschnagel, 1999].

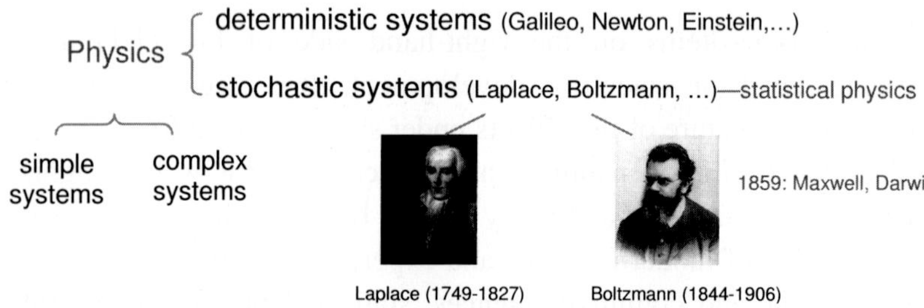

Fig. 1.1. Two ways to classify the contents of physics: (1) it is made up of deterministic systems and stochastic systems; (2) it consists of simple systems and complex systems.

Misconception 3: Science is about (mostly) simple systems only (i.e., complex systems are excluded). This misconception started from Galileo's time 400 years ago even though science before that actually includes the study of both simple and complex systems (Fig. 1.2).

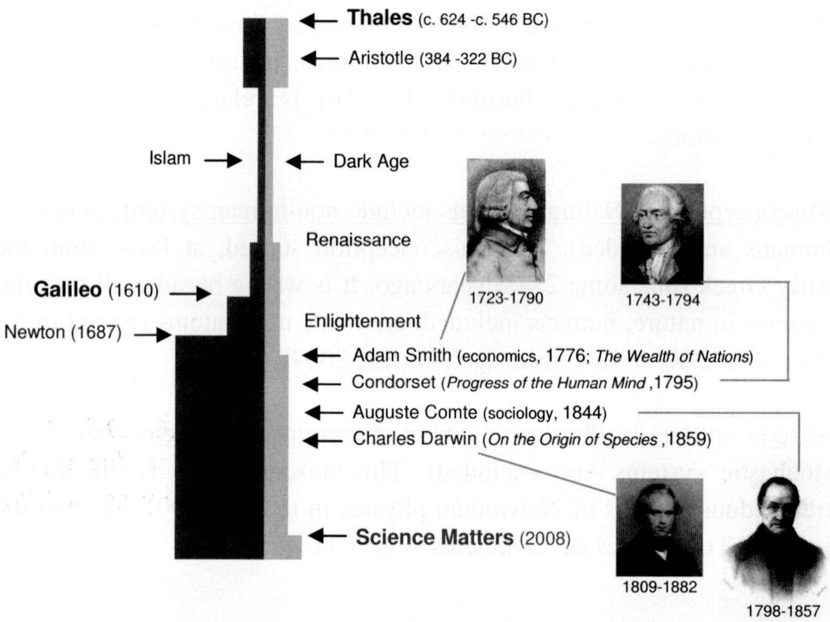

Fig. 1.2. A brief history of science in the last 2,600 years since Thales. The left (right) column corresponds to simple (complex) systems; the column width represents roughly how much the development activity was during different time periods.

1.2.3 *Science Matters*

Science Matters (SciMat) is the new discipline that treats all human-related matters, humanities in particular, as part of science (by avoiding the three misconceptions above) (Fig. 1.3). It was originated by Lam in 2008 [Lam, 2008a; 2008b]. Accordingly, Eq. (1.1) is replaced by

$$\begin{aligned}\text{science} &= \text{natural science}\\ &= \text{nonhuman-related science} + \text{Science Matters}\end{aligned} \qquad (1.2)$$

SciMat has a number of important implications [Lam, 2008a]. One of them is that the usual usage of "Science and Art" or "Science and Society" is misleading, since they imply that Science and Art (or Society) are two different things while, according to the reasoning above, art is contained within science.

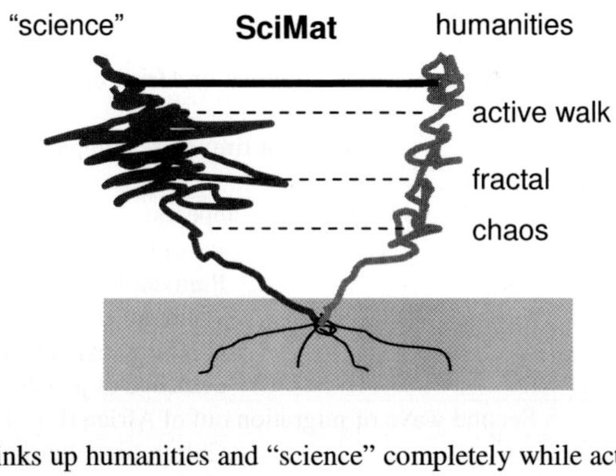

Fig. 1.3. SciMat links up humanities and "science" completely while active walk, fractal and chaos, respectively, does that partially. Humanities and "science" share the same root, growing up like two branches of the same plant.

1.3 Humans

Arts are created by humans. To understand arts we have to understand humans. Here are the basic facts about humans [Hoagland & Dodson, 1998]. A human body is composed of 5×10^{12} cells. Each cell is made up of molecules, a combination of atoms (coming from the stars). One of these molecules is the DNA molecule which is the same inside each cell. It is the DNA that passes biological information from generation to generation. However, a human being's thinking and behavior are controlled by the 10^{11} neurons in her brain. And the neurons could be influenced by external media (e.g., artwork, sunset) through the bodily sensors or substances (e.g., marijuana smoke) absorbed into the body.

There are several basic facts about human development [Mithen, 1999] that are relevant to the discussion of arts (Table 1.1).

Table 1.1. A brief history of human development. Data source: www.newscientist.com/movie/becoming-human (June 18, 2010).

Years ago	Evolution	Migration	Life style	Art related
6 million	Chimp and human lineages split.			
3.5-1.8 million			First hominids move from forest to savannah; meat eating begins.	
2.5 million	*Homo habilis* appears.			
2 million	*Homo erectus* appears.			
1.8 million		First wave of migration out of Africa begins.		
1.6 million			First use of fire; more complex stone tools created	
195,000	*Homo sapiens* (early modern humans) appears.			
400,000			Earliest evidence of cooking.	
120,000				Pigment use gives first evidence of symbolic culture.
72,000				Clothing invented and earliest evidence of jewelry
60,000		Second wave of migration out of Africa (Fig. 1.4)		
50,000				Cultural revolution: ritualistic burials, clothes-making, invention of complex hunting techniques
35,000				Oldest known cave art (in France, Spain,…)
10,000			Agriculture begins; first villages appear.	
5,500				Bronze Age begins.
5,000				Earliest known writing

Equally important is how humans pass on their genes and the question of "nature vs. nurture" [Ridley, 2003; Moore, 2003]. According to the British naturalist, Charles Darwin (1809-1882), and putting it in modern language, human inheritance is stored in the genes and passed on from generation to generation. However, *random* mutation of the genes happens from time to time, resulting in the appearance of new species. Different species compete with each other for resources and "the fittest wins." This is called "natural selection" or the evolutionary pressure; the winner keeps the (new) genes that help it win—an adaptive trait in the evolutionary sense. Moreover, the evolutionary process is very slow and continuous; no learned skills can be passed on to the next generation.

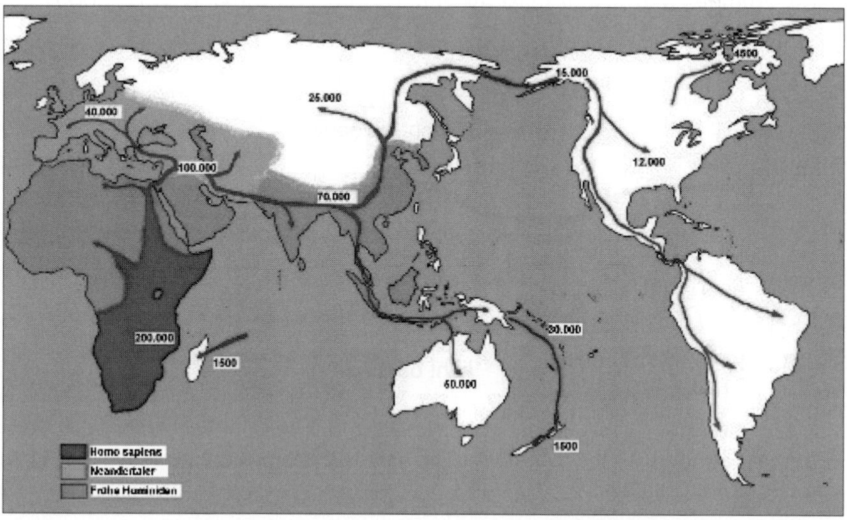

Fig. 1.4. Spreading of *Homo sapiens* out of Africa. (The numbers in this map differ slightly from those in text.) According to this map, the ancestors of the Chinese are Indians who (and everyone else) in turn are descendants of Africans.

The other school of thought was advocated by the French biologist, Jean-Baptiste Lamarck (1744–1829), who proposed that acquired characteristics can indeed be passed on to the offspring. It turns out that both Darwin and Lamarck are partially right (or partially wrong), according to new findings in genetic studies in the last decade [Jablonka & Lamb, 1995; Cloud, 2010].

The present understanding is that although we do inherit stable genes, we also inherit alterable epigenes [Shenk, 2010]. Epigenes are molecules external to the genes that can switch on and off particular genes (Fig. 1.5). More importantly, an epigene's switching state could be influenced by the environment and could be passed on to the next generation, for many generations. For example, this passing-on ability has been demonstrated in fruit flies. When exposed to a drug fruit flies show unusual outgrowths on their eyes that can last through at least 13 generations of offspring when no change in DNA has occurred [Cloud, 2010]. Similarly, experiments on roundworms fed with a kind of bacteria show changes that last at least 40 generations [Jablonka & Raz, 2009].

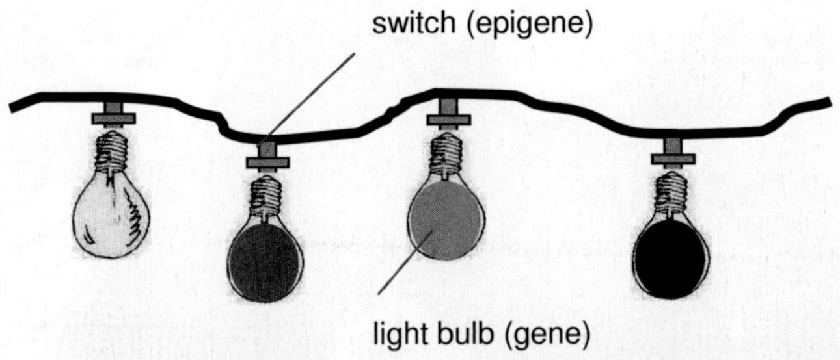

Fig. 1.5. Presently, a DNA chain can be visualized as a chain of light bulbs in different colors plus switches for the light bulbs. The light bulb represents a gene; the switch, an epigene.

What all these imply is that the debate of "nature vs. nurture" is losing its importance since both nature (genes) and nurture (environment acting through epigenes) are both inheritable, with important implication for the origin-of-arts problem (see Section 1.6).

1.4 Origin of Arts

The cave art (Fig. 1.6) 35,000 years ago [Aczel, 2009] is pretty sophisticated and mature, as painting goes, not something you and I can likely do.[4] The recent discovery of Neandertal[5] jewelry and body paint in Spain suggests that modern human behavior has ancient roots [Wong, 2010]. Even carving of 250,000 years ago had been found [Appenzeller, 1998]. Fire use (starting 1.6 million years ago) and cooking (400,000 years ago), two sophisticated inventions, occurred long before modern humans appeared. All these point to the fact that arts could start a million years ago.

No one could be sure how it happened since there is no record left from that period in time so long ago; moreover, arts like pantomime and dancing would leave no marks. Yet, since evolution happened very slowly, the basic instinct of our ancestors could not differ that much from our own. We could thus guess reasonably. And it could happen like this.

Fig. 1.6. Painting in Lascaux cave, southwestern France (c. 15,000 years ago).

[4] No cave art was found in China. For rock art in China dating to about 10,000 years ago, see [Chen, 2009].

[5] Neandertals, our closest relatives, ruled Europe for 200,000 years, but vanished about 28,000 years ago [Wong, 2009].

Let us say, somewhere in Africa, a community of few or ten people confined to a habitat during a rain of three days and nights, what would they do? And that happened after a good kill; food was plentiful and there was no rush to make preparations for the next hunt. Sex activities could use up just that much time;[6] and pretty soon, all females would be pregnant. So some of them might start doing something purely "useless," just to *kill time.*[7] Someone could be tracing another person's shadow on the soil surface with a piece of stick—early painting, hitting a piece of wood with a stick—early music, telling something with his hands and bodies—pantomime performance, and balancing the body on a piece of wood put on top of a rock—early performance art. Of course, all these could serve as entertainment, too. And it took time to perfect these skills, with more time killed. The ability to handle all these activities is "mimesis," i.e., to mime, imitate, gesture, and rehearsal of skill, which was already there 2 million years ago [Donald, 2006, p. 7]. Or, something like these might happen later, when there were tens or hundreds of persons living together in a larger group.

All these acts were not for sexual selection (which came later), because that was basically a "free sex" community. And they are never about aesthetics alone, since nature provides plenty, and better, aesthetic experiences (like the sunset).

In any case, sooner or later, someone (or a few in the group) would emerge as an expert in something and might become the first

---

[6] In a group of ten members, assume that there are two very old members and two child and we are left with eight sexually active members; further assume that four are males and four are females, and they are all heterosexual. Then there are 16 pairing of copulation; if each copulation consumes ½ hour, it will take only two to eight hours for all copulations to finish, depending on whether the copulations take place concurrently or serially.

[7] "Kill time" is a neutral word here, without value judgment; it is short for "spending one's *free* time"; it does not imply the person is bored. In modern society, let us say we work 5 days/week, 8 hours/day; and we sleep 8 hours/day and spend 1 hour traveling between home and work/day. In addition, we have to spend 2 hours/meal (including cooking, eating and washing dishes), 2 meals/day, say. We are left with $24 \times 7 - [(8 + 2 \times 2) \times 7 + (8 + 1) \times 5] = 39$ hours/week, our free time after work, eat and sleep. A million years ago and before the arts profession appeared, work consisted of hunting, gathering and tool making. How each species spends its free time differentiates one species from the other, and has important consequence in its evolutionary survival.

professional "artist," earning his food by entertaining others and staying at home instead of going out to hunt. In other words, *being an artist was the first safe job in history*. This would happen when the population expanded to a certain point and could afford to keep such an artist in their group. The first expert could be male or female, in principle.[8] The important points are:

1. Being a safe job, competition is keen and innovation was called for in the arts profession from the very beginning.
2. To keep the good job and reduce competition, the first artists would tend to maintain secrecy of their trades, pass their skills to their own sons only,[9] or inject mysterious elements into their practices—giving birth to the new profession of sorcery, perhaps.
3. The market demand and positive-feedback effect guaranteed that arts as a profession, once established, would not vanish.

As time went by and humans advanced, more varieties of arts were created. For example, with the invention of pigments, we had color painting; with language, singing; with writing, literature. And only with plenty of leisure time and a large enough market that *pure arts* appeared. Before that it was all *applied arts*,[10] which, of course, coexisted with pure arts after that. Here, pure arts mean arts for arts sake, and applied arts are done with some practical applications in mind, such as group dancing in ritual ceremonies which is to increase group adhesion.

Only in the last two hundred years that the word "art" was associated with aesthetics and fine art [Shiner, 2001]. Art in this narrow sense is part of pure arts.

---

[8] There is hint that cave arts could be made by women as well as men. See: www.dailymail.co.uk/sciencetech/article-1197680/After-25000-years-scientists-discover-artwork-created-cave-men-AND-cave-women.html##ixzz0rJjXRF3I (June 19, 2010), and [Lane, 2011].

[9] This practice is maintained today in some Asian countries, in the professions such as martial arts and Chinese medicine.

[10] Every discipline can be divided into two parts: pure and applied, like the case in physics and in history [Lam, 2002].

1.5 Nature of Arts

Arts are a subset of humans' creative activities that aim to excite the receiver's neurons in a certain manner, through that person's senses, with or without significant consequences. This is pretty strange, since while good research works in all other disciplines are also creative activities, only arts as a discipline—with the exception of entertainment—aims at someone's neurons. For instance, pure science is to understand how nature works; it does not aim at anybody's neurons; it does not even need anybody out there (apart from the creator) to receive the end results.

1.5.1 *Applied Arts*

Applied arts, by definition, are explicitly useful. For instance, a well-decorated vase will help to sell more of those vases, apart from increasing its aesthetic value when placed in a sitting room. A skillfully written novel could change the worldview of the reader, turning her into a fighter for a noble cause or a revolutionary. Obviously, architecture is one of applied arts. It is pure arts that are puzzling. What are they good for?

1.5.2 *Pure Arts*

As observed by Kant (1724-1804), art is useless [Kant, 2007]. Here, Kant is referring to pure art, and useless does not mean that it is completely void of consequences. A beautiful landscape painting, for example, could put the receiver into a serene mood. Da Vinci's *Mona Lisa* (1503-1506) could jumpstart the receiver's neurons to wonder what that lady is smiling about. Yet, apart from exercising the receiver's brain—perhaps providing pleasure to and stimulating the thinking or creativity of this person—pure arts do not seem to have any important consequences. This is not exactly true (see Criterion 2 below). Here, five criteria on *lasting* pure arts are presented.

<u>Criterion 1: Aim at receiver's neurons</u>

This is the basic characteristic of any art.

Criterion 2: Kill time

An important function of pure arts is to kill time (see footnote 7), the time of the *receiver*.[11] If it is indeed an important piece of art, it is always the case that the receiver has to spend a lot of time contemplating it, while experiencing it and *afterward*. That is what happens to Marcel Duchamp's *Fountain* (1917) (see Fig. 1.7 below), as the receiver is concerned. That is also the case for an art movie or a good play, even though the viewing time of each is about two hours only. In short, pure arts kill time on the part of the receiver; *good pure arts kill a lot of time*. And that is an important criterion on pure arts.

Criterion 3: Kill time gently and harmlessly

But "kill time" by itself is not enough for something to be called pure art. Entertainments and drugs could kill time, too. The difference is that pure arts kill time gently and harmlessly while entertainments such as a World Cup football game jerks your neurons every 10 or 15 minutes if it is good. Similarly, drug effects are usually not gentle and drug use could get you in jail. (Moreover, drugs do not satisfy Criterion 5 below.) By the same reasoning, classical music is art; heavy metal music, bordering on entertainment. My guess is that Napoleon (1769-1821) would not hang *Mona Lisa* in his bedroom, if the smile in the portrait was not that gentle.

In other words, pure arts allow us to kill time in ways that make us feel good, without exciting our neurons too violently, and thus, *encourage us to revisit them frequently*.

Criterion 4: Passivity

It is not true that people want to get involved actively in everything they do. After a day's hard work, most people would like to relax themselves passively by watching TV, for example, and, for those artistically inclined, listen to classical music or doing something, again passively.

---

[11] Ever since arts became a profession long time ago, perhaps a million years ago, killing time was not the motivation of any *professional* artist. The creative effort in arts, like that in physics, could be hard work [Lane, 2011]. As in any creative profession, the time spent is partly to make a living, but mostly it is to satisfy one's personal urge to create (and ego, for most people).

On the weekend, they might read a book or go to an art museum, enjoying arts passively.

In fact, passivity on the part of the receiver is the signature of all great arts, from painting to literature and to performance arts. All pure arts (and some applied arts) have served the receivers this way in the past many, many years, building up a habit or tradition that we humans still keep. That is why interactive arts never caught on, and perhaps will never be in the future. *Too much interactivity is bad for pure arts*.

Criterion 5: Human creation or intervention

Arts, by definition, have to be something created or intervened by humans. It does not mean that the artists cannot use materials—natural or man-made—or do their work with the aid of machines or computers. Of course, they do, and have been doing it all the time.

By this criterion, a piece of rock lying on the roadside is not a piece of art, no matter how beautiful it is. However, if you take a photo of that rock, the photo could become a piece of art—*photographic art*, because the creation of the photo involves your intervention, assuming that Criterion 1 is also satisfied (which you can help by making the photo interesting, e.g., by bettering the camera angle and using artificial lighting).[12] You might also bring that rock home, put a frame around it and become an artist instantly, because the frame is your way of telling the receiver that you want that person to look at it from a certain angle, a human intervention. Of course, there is no guarantee that this *geographic art* is a piece of great art.

As a result of the tradition a million years in the making, we treasure more those arts that are created with the less external aids. Paintings and sculptures, created practically without external aids, are high on the list. For this reason, we will never consider computer graphics (such as fractals) or ape's "painting" as high art.[13] Similarly,

[12] Such a strategy is employed by Frankel in producing images from science and engineering experiments, making the two disciplines visually informational and accessible to the public and within the research community [Frankel, 2002].

[13] One day, if computers are smart enough to create paintings or write music and novels *all* by themselves (e.g., through genetic algorithm), we may have to differentiate two

mass reproductions of an art piece are considered commercial products but not art pieces because they are too many steps away from the human creator, the artist. That explains why Andy Warhol (1928-1987) reproduced his silk-screen art pieces in a very limited number, and did it with human hands.

In summary, *pure arts are created by humans or with human intervention, to kill time gently and harmlessly, and let the receiver to experience it (preferably) passively.* With this understanding, it is obvious that the content or form of a pure-art piece is secondary;[14] they are there to serve Criteria 2 to 4.

Since the system of neurons of human beings is an extremely complex system that is not yet well understood, there are not yet sure ways to create an artwork that would satisfy Criteria 2 and 3, i.e., arousing the interest and getting the repeated attention of the receiver. All feasible ways had been tried by artists, such as appealing to humans' deep emotions about love and motherhood, and religious upbringing. However, since the brain's neuron connections are shaped not just by nature but by culture, something that worked in a previous era may not work for the present generation. What is clear is that pure arts do not always work on people's sense of aesthetics (Fig. 1.7, left); they are also about all kinds of emotions such as fear (Fig. 1.7, upper right) and other things (Fig. 1.7, lower right), too. Just like physics, arts are about the representation, description or interpretation of everything in nature. Sometimes arts abstract the real (Fig. 1.8)[15] or play on people's affinity

---

categories of arts: computer arts and human arts. Our bet is that the former will be valued far less than the latter due to Criterion 5.

[14] This does not imply that the content is immaterial. For example, in Duchamp's *Fountain*, the readymade urinal is actually a pretty complicated object with a peculiar shape that invokes all kinds of interesting thoughts, leading to the fulfillment of Criterion 2. If he replaced the urinal with a simple rice bowl, it would not work. Or, if he used a dirty urinal instead of a clean one it would not work either, because that would make the receiver uncomfortable and the resulting work would fail Criterion 3.

[15] Charlene Lam's *Petals* in Fig. 1.8 was made during a dark February in northern Sweden, which is a visualization and representation of the contrast between the region's long, light-filled days of summer and the short days of winter. The lengths of paper used were determined by the actual and predicted lengths of daylight for the first of each

for ambiguity (Fig. 1.9). Apparently, there are endless ways of doing good arts; we just do not know exactly what they are. This problem of arts is both its strength and difficulty, and obviously is an open problem in science.

Fig. 1.7. *Left*: Jean-Auguste Dominique Ingres, *The Spring* (1820-1856). *Upper right*: Edvard Munch, *The Scream* (1893-1910). *Lower right*: Marcel Duchamp, *Fountain* (1917).

---

month in 2009. The outer loops of each petal represent the 24 hours in a day; the inner loops represent the hours of daylight. It should be noted that though the length of the materials were determined by scientific data, the resulting visualization is more of an artistic statement about the perception of light available in a given day. More details and additional work can be found at www.charlenelam.com.

Fig. 1.8. Charlene Lam, *Petals (Longing for Light)* (2009), discarded paper strips and thread, 27 x 27 x 3 cm.

As humans migrated out of Africa 60,000 years ago, the contents of arts assumed local features, in addition to global universal elements developed in Africa. The fact that we treasure artworks (more than old stamps, say) implies that they do touch humans' deep emotions, needs, values, or something uniquely human. This was exemplified clearly in the French's national effort to hide the Louvre's artworks—and not something else—outside of Paris before the occupation of the Germans during World War II [Nicholas, 1995].

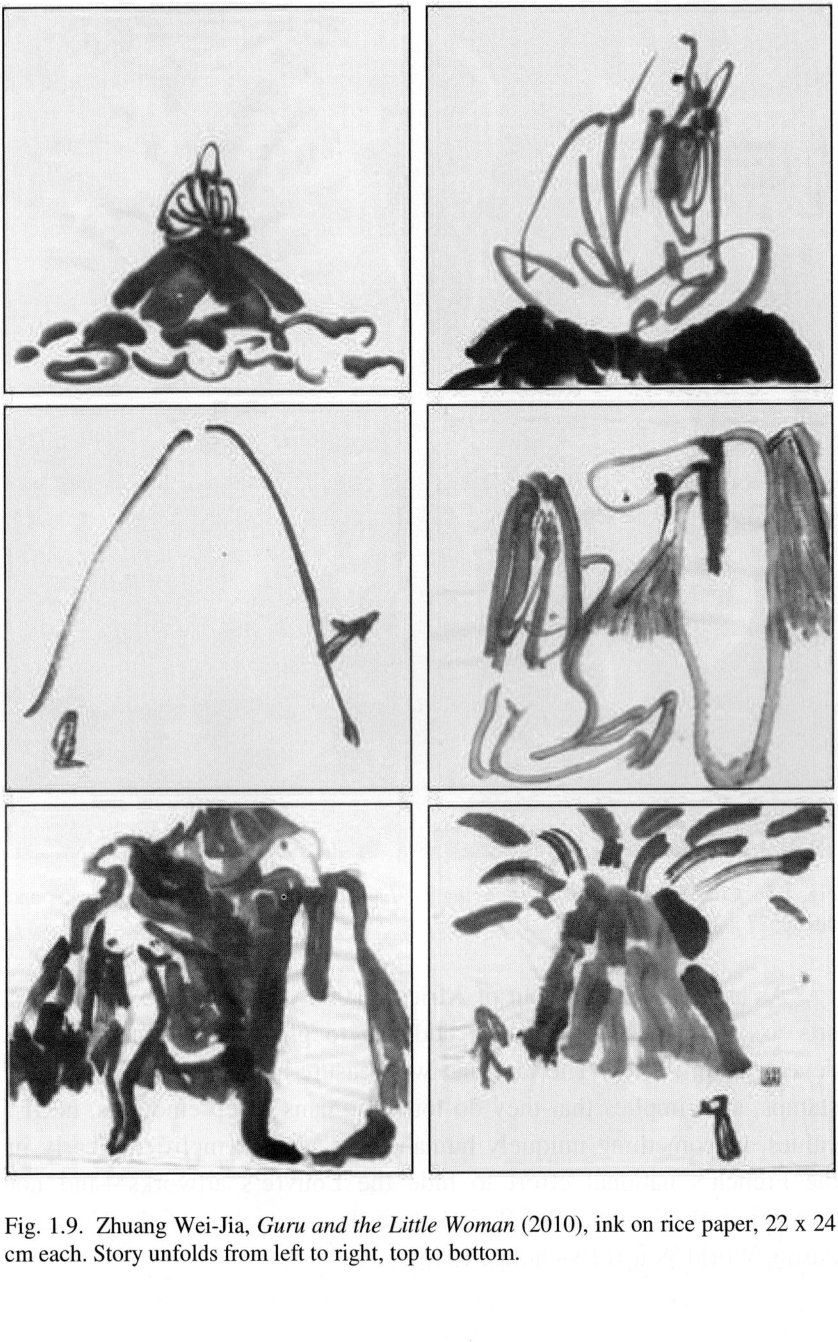

Fig. 1.9. Zhuang Wei-Jia, *Guru and the Little Woman* (2010), ink on rice paper, 22 x 24 cm each. Story unfolds from left to right, top to bottom.

1.6 Arts as a Science Matter

Arts, a part of science and a topic in SciMat, can be and should be studied scientifically. To study something is to understand it as thoroughly as possible, with all possible methods and using all appropriate tools. Therefore, knowledge and experiences from other disciplines could be borrowed; theoretical and experimental approaches are both allowed.

1.6.1 *Three Lessons from Physics*

There are three lessons that physics can offer to arts:

1. How to define a field

The domain of physics keeps on changing. For example, presently, the matured Newtonian dynamics drop out of physics and are picked up by engineering; new subdisciplines such as econophysics [Mantegna & Stanley, 2000], histophysics [Lam, 2002; 2008c] and complex systems [Lam, 1998] are added. To accommodate this ever shifting scene, *Physics Today*, the monthly magazine published by the American Physical Society, cleverly defines physics like this: "Physics is what physicists do" [Lubkin, 1998, p. 24]. Arts, with its content and form ever changing due to cultural and technological advancements, are like physics in this regard. It is thus not surprising that efforts to find the necessary and sufficient definition of arts all fail [Carroll, 1999].[16]

2. The existence of subdisciplines

In physics, different aspects of the same material are studied in different subdisciplines. As an example of the latter, for solids, we have Optics of Solids, Mechanics of Solids, Thermodynamics of Solids, etc. Every such subdiscipline can describe only certain properties of solids. They complement each other; when combined, a full understanding of solids is

---

[16] Noël Carroll's practical reason [1999, p. 207] that we need to define art precisely so an artwork such as Brancusi's abstract sculpture *Bird in Flight* can be imported to the U.S. duty free, is not strictly valid. This practical problem can be solved in the same way that we handle pornography, i.e., to be decided by a committee of local residents.

achieved. When viewed this way, the different approaches in art studies (such as Marxism, feminism, biography and autobiography, semiotics, psychoanalysis and aesthetics) [Adams, 1996] are actually studies at the phenomenological level (see below); each approach is a subdiscipline that specializes on a narrow aspect of arts. No one should expect that any one of them to be encompassing and be able to tell the whole story.

3. <u>There are three levels of study</u>

In *any* scientific study, after observing and collecting data, and analyzing the data, there are three approaches or levels—empirical, phenomenological and bottom-up—that one can adopt to go further [Lam, 2002]. These three approaches in the cases of physics and arts are sketched in Table 1.2. Empirical studies always happen first. Phenomenological studies are done without knowing the mechanism underlying a phenomenon; they are very powerful and sometimes undervalued. Fundamental understanding of a phenomenon is reached through the bottom-up studies in which the mechanism will reveal itself and become understood.

Table 1.2. The three approaches in the study of physics (gas as an example) and arts.

Approach	Gas	Arts
Empirical	Gas law	Empirical rules discovered by artists; empirical studies by art critics/historians/ scholars; fractal analysis of paintings and music by physicists
Phenomenological	Navier-Stokes equation	Interpretations of nature of art by art philosophers/ historians/scholars; evolutionary arguments via Darwin
Bottom-up	Molecular picture (called microscopic method)	Studies through biology: evolution theory (genetics), cognitive science, neuroscience; through physics: statistical analysis

1.6.2 *Arts Studies in Three Approaches*

Here examples of arts studies done with the three approaches are given.

1. <u>Empirical</u>

In arts studies, at the empirical level, there are various empirical rules worked out by the artists and empirical analyses due to arts scholars. An example of the latter is the work by André Leroi-Gourhan (1911-1986); he regarded *all* the signs *and* animals depicted in the prehistoric cave paintings as sex symbols and classified them as either female or male (e.g., bison for female, horse for male), reflecting the supposed worldview that the world is divided into two types or two kinds of things, two genders, akin to the Chinese's yin-yang philosophy (see [Aczel, 2009]). Another example is Taylor's fractal analysis [2002] of the dripping paintings of Jackson Pollock (1912-1956), and the fractal studies of music (see [Barrow, 1995]). The third example is the statistical analyses of literature by Gottschall and his coworkers [Gottschall, 2008]; e.g., romantic love is shown to be a literary universal by counting the number it occurs in folktales from different cultures.

2. Phenomenological

At the phenomenological level, apart from the different interpretations of arts such as those summarized in Adams' book [1996] mentioned above, the origin of arts was studied by Dissanayake [1988] through Darwin's evolution theory. Here, arts are argued to be an adaptive behavior that benefits humans' survival, passing from generation to the next through the genes. The problem with this approach is that, in view of the new findings in epigenetics (Section 1.3), it is not clear that the genetic route is at all necessary. Maybe the cultural effects on arts could be inherited too, through the epigenes; or both. The crucial problem is that unless there is experimental proof of the adaptive nature of arts, such a conjecture remains a conjecture. Similar considerations plague the debate of the nature vs. culture origin of arts. Further examples of the phenomenological approach are the evolutionary study of literature by Carroll [2004] and of film by Anderson [1996].

3. Bottom-Up

The bottom-up approach in the study of physics of materials is to start from the molecular description and work out the macroscopic properties; that is because molecules are at the lower level of materials. The case in humanities is more complicated and slightly different. Even though the immediate lower level of a human body is consisting of cells, no one ever tried to understand humans by considering trillions of cells together.

Instead, the bottom-up approach in the humanities starts from a much lower level, from either the neurons or the genes. The former leads to *neurohumanities*, with subdisciplines such as neurophilosophy and neurotheology, and, for arts, neuroesthetics [Skov & Vartanian, 2009], neuromusicology, neuroarthistory [Onians, 2007] and neurocinematics [Hasson et al, 2008]; this rapid development of neuro-based studies is sometimes called the Neuro Revolution [Lynch, 2009]. The latter leads to the genetic study of humanities and is less developed.

That neuroscience and cognitive science [Turner, 2006] are important in arts studies is not at all surprising, since the brains of the artists and the receivers are obviously important in the creation and appreciation of artworks. And neuroscience helps us to understand what makes the human brain unique [Gazzaniga, 2008]. Yet, in the neuro-studies of arts, some progresses are made but success is limited, perhaps due to the fact that the development in neuroscience and cognitive science [Kolak et al, 2006] is still in its infancy. More nontrivial results are needed to get the attention of the artists.[17]

Apart from the books mentioned above, neuroscience in connection with literary studies is discussed in [Hogan, 2003a; 2003b]. Applications of neuroscience in movie (and advertisement) studies are reported in [Hamzelou, 2010]. Examples of specific works connected with global art history [Onians, 2011] and the paintings of Su Dong-Po and Paul Cézanne [Lam & Qiu, 2011] as well as other topics can be found in the book *Arts: A Science Matter* [Burguete & Lam, 2011].

1.7 Arts and "Science"

The issue of arts and "science" (see footnote 1) is of great interest to many people. What this actually means is that whether there is any connection between the arts and the study of the physical world, since both are creative processes. Unfortunately, there exist a lot of confusion concerning this issue, due mainly to the misunderstanding of the domain

[17] In physics, e.g., the nontrivial result that maximum range of a projectile is achieved with a throwing angle of 45° is taught in high school. This knowledge is used by the teenager in the movie *Aliens in the Attic* (2009) in hitting the alien with a dart and thus saving the world!

of science, and the origin and nature of arts. Our takes on this issue are summarized in Table 1.3.

Table 1.3. Comparison between arts and "science."

Characteristics	Arts	"Science"
Both are part of science	Arts are part of science.	"Science" (mostly about nonliving systems) is also part of science.
Different aims	Arts aim at receiver's neurons.	"Science" aims to understand how nature works.
Receiver	Arts need a receiver to appreciate the artwork.	"Science" needs no receiver (but has to compare with nature, the ultimate judge).
Different history	Arts started at least 35,000 (and could be a million) years ago.	"Science" started about 2,600 years ago since Thales (c. 624-c. 546 BC), after the invention of language and writing.
Relationship between arts and "science"	• Both involve creative process (for different reasons)—but same in many other human activities. • Arts are humans' creation, reflecting on the world of human and nonhuman systems; the principles governing this world are the same principles (e.g., symmetry, spontaneous symmetry breaking, fractal, chao, active walk) studied by "scientists." • Progress in "science" (and related technology) advances the development of arts; e.g., pigments→color painting, film/camera→photographic art, electricity→cinema, laser→photon art, computer→digital art.	

Furthermore,

1. There is a common misunderstanding on how artists and "scientists" use their brains. The truth is that these two kinds of professionals both use intuitions and rational thinking in their trades [Lam, 2004]; they use both their left and right brains. Real-time brain scans using fMRI (functional Magnetic Resonance Imaging, started in 1990) could help to clarify this issue.
2. Since artists, painters in particular, quite often depict the world as it appears to them, the same physical principles that govern the

working of the physical world could show up in the artists' works. This is indeed the case; some of these principles such as symmetry and broken symmetry are discussed in [Leibowitz, 2008].

3. Occasionally, artists are inspired by "science" in creating their work (see Fig. 1.10 and [Lam, 1998, p. 19; Burguete, 2011]). The reverse case is discussed in [Burguete, 2011].
4. For artists and scientists living in the same era, will their works influence each other? There is no direct evidence that this happened, but they, like Picasso and Einstein, certainly were aware of each other's work [Miller, 1996; 2001].
5. Since artists are those with supreme bodily sensors [Lam, 2008a] and using them everyday, they could know something empirically on how the brain works, long before the "scientists" find it out in their labs. Examples on this are given in [Strosberg, 2001; Lehrer, 2008]; see also [Gardner, 1993].
6. Examples are given by Edwards [2008] that some people can cross the boundary between art and "science," and achieve breakthroughs in both. Discussions on the link between "science" and culture are given in [Slingerland, 2008; Galison et al, 2001].
7. Selected artists and "scientists" like to present their effort as the search for "truth, virtue and beauty." But this claim could not be universally true, for the following reasons: (1) Truth is a fuzzy concept [Godfrey-Smith, 2003]. (2) No virtue could be found by naturalists in their study of how insect parasites feed on their hosts;[18] virtue is claimed mostly by theoretical physicists. (3) As shown in Fig. 1.7, arts are not always about "beauty"; some physicists found Newton's second law of motion, $F = ma$, beautiful (or elegant), but that is because a lot of messy details are hidden from this expression.[19]

---

[18] For example, predatory wasps paralyze the host insect by stinging and lay eggs on it. The larvae feed on the paralyzed host and killing it in the end (http://en.wikipedia.org/wiki/Entomophagous_parasite, Nov. 1, 2010).

[19] For instance, this law is about non-existent point particles which have mass but zero size; $F = ma$ is only a special case (when m = const) of the second law which states that $\boldsymbol{F} = \mathrm{d}(m\boldsymbol{v})/\mathrm{d}t$, a vector equation involving differential calculus; two quantities, force F and

8. Some arts are known to have healing effects [McCaffrey, 2007]; it links arts to medical science.

Further discussions on the issue of arts and "science" are provided in [Jaroszynski & McNeil, 2000; Rodgers, 2002; Shahn, 2002; Crease, 2003; Hidetoshi & Rothman, 2008; Lucibella, 2010], and references therein.

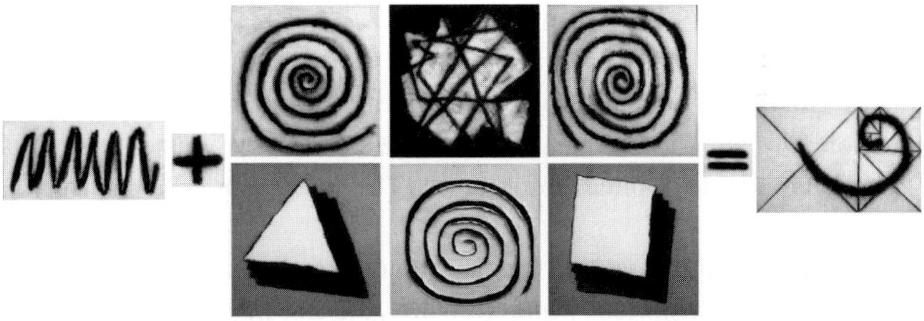

Fig. 1.10. George Cladis, *Line Revisited*, 10 acrylic and stitching on linen, 146" x 472" installation. The complete absence of color in this installation idealizes the heft and vastness of the conscious domain (for more information, see "Patterns of Mind" in http://alct.com/gCladis/AbstractCD/2abs.htm).

1.8 Discussion and Conclusion

The nature and origin of arts is a 2,400-years-old unsolved puzzle dating to Plato's time. In this chapter, an answer to this puzzle is attempted.

In previous sections, we point out that arts consist of applied arts and pure arts. Applied arts could start a million of years ago, while pure arts emerged later when the community is large enough to support pure artists. Being an artist was the first safe job in human history. The nature of arts is then discussed; five criteria on pure arts are given. Our understanding of arts is based on a global and historical perspective; it differs from the conventional wisdom popularized by the West [Shiner, 2001, p. 6] or existing in the East. Yet, reasonable this story of arts is, it has to be substantiated by more works. Here are further discussions.

---

mass *m*, are defined with this one equation, which is completely illogical [Wilzeck, 2004].

1. In a way, arts are like driving. Driving is to move a vehicle from point A to point B and deliver something. The motivation of the driver is irrelevant here; the type of vehicle used is secondary. Similarly, the essence of arts is to have the artwork created in the artist's brain (point A) and delivered to the receiver's neurons (point B). It will only confuse the issue if you search for the artist's motivation (which could be a mixture of curiosity, fame and making a living—same as for scientists [Tsui & Lam, 2011]).[20] And the content and style of the artwork, like the vehicle in driving, is secondary as long as they sit well with the receiver's neurons.

2. Historically, the development of arts is from applied to pure (Section 1.4), similar to the case in science's early development which is from technology to applied and to pure. (The reverse order is mostly true in science today.)

3. The key to understand arts is to understand arts (and not just art) as a whole, both historically and globally, and not merely the small part of it existing in Europe. In fact, the latter has a short history of only two thousand years or so, including the last two hundred years of fine arts [Hoppe, 2011].

4. It seems that women fall easily for male artists, but female artists do not attract men, at least not to the same degree. If this is true, it can be explained by our surmise that arts was the first safe job in prehistory. According to the evolution theory, women tend to choose capable men who would and could stay around to protect the mother and child. In this respect, an artist father would be favored over a hunter father, and this preference could be carried over to the present day. More research on this issue is warranted.

5. In the West, arts (and "science") were derived from the "liberal arts" some 2,000 years ago and the ideal of liberal arts—designed for free men—was the pursuit of universal knowledge [Wu, 2011]. On the other hand, artworks in Europe until a few hundred years ago was operated within the commission/patronage system; the artist was not

[20] Einstein's explanation for people's motivation in doing art and science is that "[to] escape from everyday life with its painful crudity and hopeless dreariness" [Einstein, 1982, p. 225].

completely free in choosing his topics, not even Michelangelo or da Vince [Shiner, 2001, p. 18]. Surprisingly, great arts were produced. There were two factors that made this possible: (1) the artist was given enough artistic freedom in executing the project, and, more importantly, (2) the person who had the final say, the patron, possessed a high enough level of sophistication or artistic taste. That the artistic sophistication of the "authority" is crucial is reflected in the uniform style of the tall buildings built in Beijing, China, within a couple of years about a decade ago. They all wear a "hat"—the Chinese way of blending the Western and Chinese styles of architecture at that time; an example is shown in Fig. 1.11. Even in the last two hundred years when there has been a free market for artworks, in a certain sense, artists are still not completely free. The art creating process is constrained by the human nature of the artist and the receiver, and by the principles underlying nature.

Fig. 1.11. Pacific Digital City, a "hat"-wearing building at Peking University, Beijing, China.

6. Neuroscience in arts studies could involve the use of fMRI scan and optogenetics [Deisseroth, 2010], applied to the brains of the artist and the receiver. In the optogenetic experiments, light-responsive opsin genes are inserted into the cells of the brain. Specific neurons can then be triggered to fire by a flash of light, providing precision down to the neuron level.

7. In arts studies, no matter how useful the bottom-up neuro approach (theoretically and experimentally) is, it should not be the only method one relies on in understanding arts. Empirical and phenomenological approaches as well as *common sense* should not be forgotten. The situation is like that in physics. After quantum mechanics was invented about 100 years ago, there was a rush to do quantum theory of everything interesting. However, if one wants to understand the flow of water, the advice is to start from the phenomenological Navier-Stokes equation and not quantum mechanics, because it is unnecessary and impractical to do so. In the study of complex systems (and in any other discipline), the appropriate research tool should be selected according to the level at which the property under study emerges [Lam, 2002].

8. In [Dutton, 2009, pp. 52-59], 12 features of art are listed: direct pleasure, skill and virtuosity, style, novelty and creativity, criticism, representation, special focus, expressive individuality, emotional saturation, intellectual challenge, art traditions and institutions, and imaginative experience. Unfortunately, it seems that all these features are shared by the other creative discipline called physics, and hence are not unique to art.

9. The monetary value of an art piece fluctuates in time. It has more to do with the uniqueness of the artwork, like a large piece of diamond or rare stamps, than to do with the artistic nature of the artwork itself.

10. Religion and "science" have been called the two pillars of Western civilization by the Nobel laureate in physics, Richard Feynman (1918-1988).[21] Since religion is not a prime element in societies like China, it could be said that *it is arts and "science" that are the two*

---

[21] "The Relation of Science and Religion," the transcript of a talk given by Richard Feynman at the Caltech YMCA Lunch Forum on May 2, 1956 (http://calteches.library.caltech.edu/49/2/Religion.htm, June 18, 2010).

pillars of any *civilization*. That is, arts are more basic than religion as civilization is concerned.

11. In principle, all modern creations by humans—like cell phones and air conditioning—could be given up and (most if not all) humans would still survive. However, without these creations, life would become very inconvenient and many people would suffer. In contrast, if *Mona Lisa* was gone, only the Louvre museum would suffer, materially speaking. In this sense, and only in this sense, arts are inessential and "useless." Otherwise, arts are most valuable! No one should live in a world without arts, and they can't.

References

Aczel, A. D. [2009] *The Cave and the Cathedral* (Wiley, Hoboken, NJ).

Adams, L. S. [1996] *The Methodologies of Art: An Introduction* (Westview, Bouler, CO).

Anderson, J. D. [1996] *The Reality of Illusion: An Ecological Approach to Cognitive Film Theory* (Southern Illinois University Press, Carbondale and Edwardsville).

Appenzeller, T. [1998] "Art: Evolution or revolution?" *Science* **282**, 1451.

Barrow, J. D. [1995] *The Artful Universe: The Cosmic Source of Human Creativity* (Little, Brown and Co., New York).

Brown, S. & Dissanayake, E. [2009] "The arts are more than aesthetics: Neuroaesthetics as narrow aesthetics," in *Neuroaesthetics*, eds. Skov, M. & Vartanian, O. (Baywood, Amityvile, NY) pp. 43-57.

Burguete, M. [2011] "ChemArt and BioArt: Art-science interactions," in *Arts: A Science Matter*, eds. Burguete, M. & Lam, L. (World Scientific, Singapore).

Burguete, M. & Lam, L. [2011] *Arts: A Science Matter*, eds. Burguete, M. & Lam, L. (World Scientific, Singapore).

Carroll, J. [2004] *Literary Darwinism: Evolution, Human Nature, and Literature* (Routledge, New York).

Carroll, N. [1999] *Philosophy of Art: A Contemporary Introduction* (Routledge, New York).

Chen, Zhao-Fu [2009] *Discovery History of Rock Art in China* (Shanghai People's Publishing House, Shanghai).

Cloud, J. [2010] "Why your DNA isn't your destiny," *Time*, Jan. 6.

Crease, R. P. [2003] "The Newton-Beethoven analogy," *Phys. World*, April, 16.

Deisseroth, K. [2010] "Controlling the brain with light," *Sci. Am.*, Nov., 49-55.

Dissanayake, E. [1988] *What Is Art For?* (University of Washington Press, Seattle).

Donald, M. [2006] "Art and cognitive evolution," in *The Artful Mind: Cognitive Science and the Riddle of Human Creativity*, ed. Turner, M. (Oxford University Press, Oxford).

Dutton, D. [2009] *The Art Instinct: Beauty, Pleasure, and Human Evolution* (Bloomsbury, New York).

Edwards, D. [2008] *Artscience: Creativity in the Post-Google Generation* (Harvard University Press, Cambridge, MA).

Einstein, A. [1982] *Ideas and Opinions* (Three Rivers Press, New York).

Frankel, F. [2002] *Envisioning Science: The Design and Craft of the Science Image* (MIT Press, Cambridge, MA).

Galison, P., Graubard, S. R. & Mendelsohn, E. [2001] *Science in Culture* (Transaction Publishers, New Brunswick, NJ).

Gardner, H. [1993] *Creating Minds: An Anatomy of Creativity Seen Through the Lives of Freud, Einstein, Picasso, Stravinsky, Eliot, Graham and Gandhi* (Basic Books, New York).

Gazzaniga, M. [2008] *Human: The Science Behind What Makes Us Unique* (HarperCollins, New York).

Godfrey-Smith, P. [2003] *Theory and Reality: An Introduction to the Philosophy of Science* (University of Chicago Press, Chicago).

Gottschall, J. [2008] *Literature, Science, and a New Humanities* (Palgrave MacMillan, New York).

Hamzelou, J. [2010] "Brain invasion at the multiplex," *New Scientist*, Sept. 11, 8-9.

Hasson, U., Landesman, O., Knappmeyer, B., Vallines, I., Rubin, N. & Heeger, D. J. [2008] "Neurocinetmatics: The Neuroscience of film," *Projections* **2**(1), 1-26.

Hidetoshi, F. & Rothman, T. [2008] *Sacred Mathematics: Japanese Temple Geometry* (Princeton University Press, Princeton).

Hoagland, M. & Dodson, B. [1998] *The Way Life Works* (Times Books, New York).

Hogan, P. C. [2003a] *Cognitive Science, Literature, and the Arts: A Guide for Humanists* (Routledge, New York).

Hogan, P. C. [2003b] *The Mind and Its Stories: Narrative Universals and Human Emotion* (Cambridge University Press, Cambridge, UK).

Hoppe, B. [2011] "The Latin *Artes* and the origin of modern *Arts*," in *Arts: A Science Matter*, eds. Burguete, M. & Lam, L. (World Scientific, Singapore).

Jablonka E. & Lamb, M. J. [1995] *Epigenetic Inheritance and Evolution: The Lamarckian Dimension* (Oxford University Press, Oxford).

Jablonka, E. & Raz, G. [2009] "Transgenerational epigenetic inheritance: Prevalence, mechanisms, and implications for the study of heredity and evolution," *Quarterly Review of Biology* **84**, 131-176.

Jaroszynski, D. & McNeil, B. [2000] "Science given an artistic licence," *Phys. World*, July, 51.

Kant, I. [2007] *Critique of Judgement* (Oxford University Press, Oxford).
Kolak, D., Hirstein, W., Mandik, P. & Waskan, J. [2006] *Cognitive Science: An Introduction to Mind and Brain* (Routledge, New York).
Lam, L (ed.) [1997] *Introduction to Nonlinear Physics* (Springer, New York).
Lam, L. [1998] *Nonlinear Physics for Beginners: Fractals, Chaos, Solitons, Pattern Formation, Cellular Automata and Complex Systems* (World Scientific, Singapore).
Lam, L. [2002] "Histophysics: A new discipline," *Mod. Phys. Lett. B* **16**, 1163-1176.
Lam, L. [2004] *This Pale Blue Dot: Science, History, God* (Tamkang University Press, Tamsui).
Lam, L. [2008a] "Science Matters: A unified perspective," in *Science Matters: Humanities as Complex Systems*, eds. Burguete, M. & Lam, L. (World Scientific, Singapore).
Lam, L. [2008b] "Science Matters: The newest and biggest interdicipline," in *China Interdisciplinary Science*, Vol. 2, ed. Liu, Z.-L. (Science Press, Beijing).
Lam, L. [2008c] "Human history: A Science Matter," in *Science Matters: Humanities as Complex Systems*, eds. Burguete, M. & Lam, L. (World Scientific, Singapore).
Lam, L. & Qiu, Li-Meng [2011] "Su Dong-Po's bamboo and Paul Cézanne's apple," in *Arts: A Science Matter*, eds. Burguete, M. & Lam, L. (World Scientific, Singapore).
Lane, H. [2011] "From curiosity to creation: The art of Holly Lane," in *Arts: A Science Matter*, eds. Burguete, M. & Lam, L. (World Scientific, Singapore).
Lehrer, J. [2008] *Proust Was a Neuroscientist* (Mariner, New York).
Leibowitz, J. R. [2008] *Hidden Harmony: The Connected Worlds of Physics and Art* (Johns Hopkins University Press, Baltimore).
Lubkin, G. B. [1998] "A personal look back at *Physics Today*," *Phys. Today*, May, 24-29.
Lucibella, M. [2010] "Physics stars in theater, music and dance," *APS News*, Nov., 1.
Lynch, Z. [2009] *The Neuro Revolution: How Brain Science Is Changing Our World* (St. Martin, New York).
Mantegna, R. N. & Stanley, H. G. [2000] *An Introduction to Econophysics* (Cambridge University Press, New York).
McCaffrey, R. [2007] "The effect of healing gardens and art therapy on older adults with mild to moderate depression," *Holist Nurs. Pract.* **21**(2), 79-84.
Miller, A. I. [1996] *Insights of Genius: Imagery and Creativity in Science and Art* (Copernicus, New York).
Miller, A. I. [2001] *Einstein, Picasso: Space, Time and the Beauty That Causes Havoc* (Basic Books, New York).

Mithen, S. [1999] *The Prehistory of the Mind: The Cognitive Origins of Art and Science* (Thames and Hudson, New York).

Moore, D. S. [2003] *The Dependent Gene: The Fallacy of "Nature vs. Nurture"* (Freeman, New York).

Nicholas, L. H. [1995] *The Rape of Europa: The Fate of Europe's Treasures in the Third Reich and the Second World War* (Vintage Books, New York).

Onians, J. [2007] *Neuroarthistory: From Aristotle and Pliny to Baxandall and Zeki* (Yale University Press, New Haven).

Onians, J. [2011] "Neuroarthistory: Reuniting ancient traditions in a new scientific approach to the understanding of art," in *Arts: A Science Matter*, eds. Burguete, M. & Lam, L. (World Scientific, Singapore).

Paul, W. & Baschnagel, J. [1999] *Stochastic Processes: From Physics to Finance* (Springer, New York).

Ridley, M. [2003] *The Agile Gene: How Nature Turns on Nurture* (Perennial, New York).

Rodgers, P. [2002] "Physics meets art and literature," *Phys. World*, Nov., 29 (special issue on physics and arts).

Shahn, E. [2002] "Swept into the modern along with science," *Science* **298**, 2333-2334.

Shenk, D. [2010] *The Genius in All of Us: Why Everything You've Been Told About Genetics, Talent, and IQ Is Wrong* (Doubleday, New York).

Shiner, L. [2001] *The Invention of Art: A Cultural History* (University of Chicago Press, Chicago).

Skov, M. & Vartanian, O. (eds.) [2009] *Neuroaesthetics* (Baywood, Amityville, NY).

Slingerland, E. [2008] *What Science Offers the Humanities: Integrating Body and Culture* (Cambridge University Press, Cambridge, UK).

Strosberg, E. [2001] *Art and Science* (Abbeville Press, New York).

Taylor, R. P. [2002] "Order in Pollock's chaos," *Sci. Am.*, Dec., 116-121.

Tsui, Hark & Lam, L. [2011] "Making movies and making physics," in *Arts: A Science Matter*, eds. Burguete, M. & Lam, L. (World Scientific, Singapore).

Turner, M. (ed.) [2006] *The Artful Mind: Cognitive Science and the Riddle of Human Creativity* (Oxford University Press, New York).

Turner, M. S. [2009] "The universe," *Sci. Am.*, Sept., 36-43.

Wilczek, F. [2004] "Whence the force of $F = ma$? I: Culture shock," *Phys. Today*, Oct., 11-12. [Parts II and III appear, respectively, in Dec., 2004 and July, 2005 issues of *Phys. Today*.]

Wong, K. [2009] "Twilight of the Neandertals," *Sci. Am.*, Aug., 32-37.

Wong, K. [2010] "Did Neandertals think like us?" *Sci. Am.*, June, 72-75.

Wu, Guo-Sheng [2011] "Science and art: A philosophical perspective," in *Arts: A Science Matter*, eds. Burguete, M. & Lam, L. (World Scientific, Singapore).

PART I

Philosophy and History of Arts

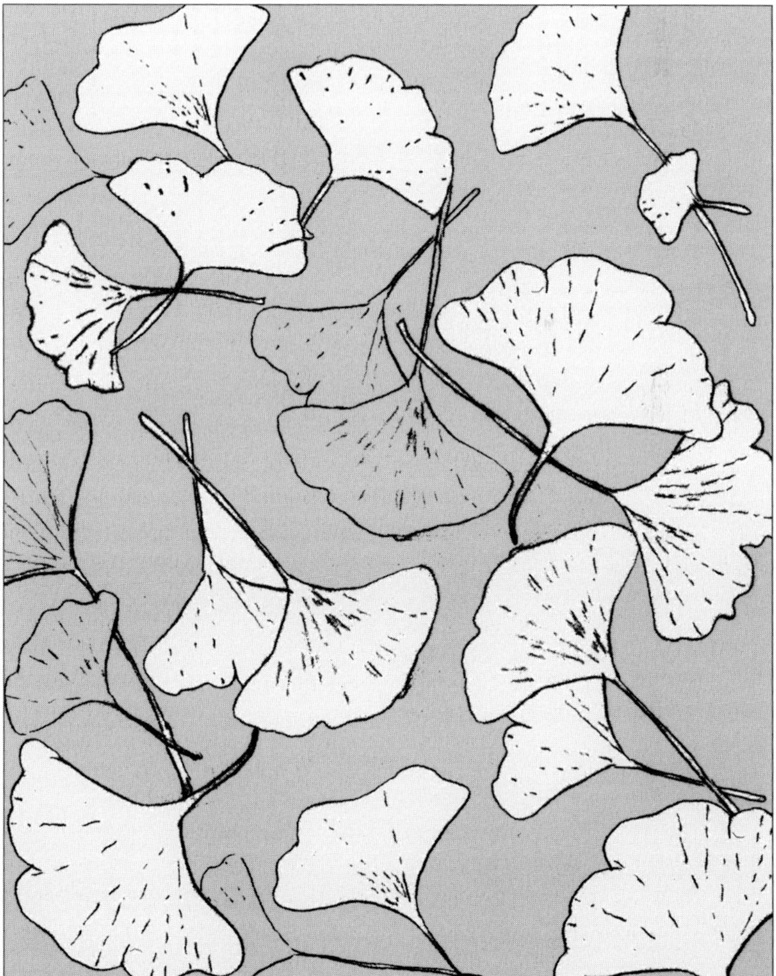

2

The Latin "*Artes*" and the Origin of Modern "Arts"

Brigitte Hoppe

This chapter will analyze the range of attitudes held by scientists and artists observing and presenting natural phenomena at various periods of time. It will explain the meaning and significance of the principal fields of education called the "liberal arts" and the "mechanical arts" since the Late Antiquity. Both areas comprised no less than seven disciplines including sciences, mathematics, artistic and technical skills. In the European civilization, these were the essential elements of education and learning until the 18th century. Certain interrelations between sciences and arts can be better understood by examining particular early collections of artifacts produced in Early Modern Times (16th–17th centuries). Sciences and arts began to deviate from one another in the course of growing specialization and changing societies in the modern world. In Antiquity (4th BC–5th centuries AD) and from the Late Middle Ages (15th century) on, artists and scientists became eminent observers of natural objects such as plants, animals, minerals, the stars in the sky, and the topography of special regions, e.g., the Netherlands coastal areas with its sailing ships, and the tropical forests of Brazil. Portraits of scientists often include typical objects such as a skull representing the field of anatomy, or an armillary sphere for astronomy. Since the 17th century we find critical discussions of scientific life, work, and the impact they made: the focused concentration of an alchemist and the social impact of scientific innovations. The discussion of the relationship between technical inventions, e.g., the steam engine, cars on a highway, multi-storey buildings in a small city street, and the bare necessities of human life furthered the development of a critical worldview in fine arts. In abstract sculptures and paintings, in particular in the style of "concrete art", geometric figures and spectral colors recur, but their meaning in an individual composition is different from a direct representation; artists of our time are able to present a unique interpretation of the world around them in their works of individual imagination.

2.1 Introduction

The history of the arts, in particular of fine arts, from the earliest periods of human civilization to the works of the present has experienced many aims and intentions. Historians of the arts and the sciences have taken full note of the many scientific motifs found in works of the fine arts, e.g., the alchemist in his laboratory [Burland, 1988]. They have also looked closely at the very detailed studies of landscapes and of natural objects, such as plants, animals, and minerals, by famous artists like Albrecht Dürer [Salley, 2003], in particular during the Renaissance. With regard to themes, the relationship between arts and sciences is well known. When examining the early development of the sciences, we find that their origin as disciplines in European civilization had its roots in a special "canon" of subjects of learning which was the basis of education starting in the Late Antiquity (5th century AD). The main aspects of that development, including the range of meaning and significance of the Latin notions of different kinds of "*artes*", were handled in the history of education and learning in fairly great detail over many years [Meyer, 1885-1887; Bernt, 1980; Christes, 1997; Lindgren, 2004].

However, the interrelationships between the development of the history of education and learning, including the early history of sciences on the one hand, and of the history of the arts on the other, were rarely discussed in detail. Even treatises on the history of different kinds of early collections, such as cabinets of curiosities tended to disregard the details of that relationship [Grote, 1994; Mauriès, 2002; MacGregor, 2007] with a few exceptions (in this regard, some remarks are included in [Klemm, 1973; Hoppe, 1994; Roth, 2000]).

From the standpoint of the history of science we may ask: was the representation of scientific motifs in the arts the only and the most important aspect of the relation between sciences and the arts? And furthermore, we are asking, regarding the interest in special scientific motifs: was it the result of free individual choice and the preference of the artist alone, or was the artist and were the clients and sponsors of his work influenced by dominant cultural tendencies during different periods? Did these tendencies have an impact on the interpretation of natural phenomena observed by scientists and artists as well? Did the attitudes of artists and scientists toward nature remain the same from the earlier historical periods to Modern Times, or did they evolve? If so, how may these changes throughout history being characterized?

By discussing the relationship between the disciplines of education called "*artes*" in the European cultural tradition and human activities which we call "*arts*" in many European languages, it is elucidating to analyze the etymology and the meaning of the word. In order to follow the development and important changes in the perception of natural phenomena by comparing the objects used and the results obtained by scientists with the works of artists, we will elaborate on the system of knowledge known in Latin as the "*artes*". Then we will analyze its primary relationship to objects of art and discuss, in particular, the role it played in the structure of the "cabinet of curiosities" and the significance of this type of collection to artistic works in Early Modern Times, keeping in mind certain significant changes in the methodology and results of modern sciences. To conclude, we will investigate similar changes and how they are reflected in the arts, especially in works of the fine arts from the 19th century to the present.

2.2 Meaning and Etymology of the Word "Art"

The word "art" is currently used in Romance and Anglo-Saxon languages to designate the object created by an artist, i.e., of a human being with a special skill not common to all humans. The etymological roots are found in the Latin word *ars* (*-tis*; with the noun and adjective *artifex*, *-ficis*) [Georges, 1879] and in the Greek word *artýō* or *artýnō*, which means I arrange, I know how to arrange, I join, and I contrive. The Greek noun *hē aretē*, meaning the ability, faculty, virtue or knowledge, is related in meaning [Liddell et al, 1958].

In like manner, the word *art* in modern languages means the forming and fashioning or modification of objects (mainly natural ones in earlier times) by human skill, in order to prepare them for a purpose intended by humans. In Classical Antiquity (4th century BC), the ability to fashion a material object like a chair or the faculty to create an artificial product like a sculpture was denoted by the word *hē téchnē*. Aristotle introduced a remarkable distinction: in his system of natural philosophy, a technical process, such as joinery or carpentry or architecture, wrought by human hand, was the opposite of a natural or physiological one. The "maker" acts, as the Greek philosopher supposed, in opposition to nature, *contra naturam*, in Latin. Aristotle had respect for all sorts of natural phenomena and humans observing them and thinking about them. According to his metaphysical distinction, the work of an artist produced by the aid of a manual skill and from non-living material had a lower value. Whereas until the 17th century the social standing of an artist was equivalent to that of a workman or an artisan, in contemporary times it ranks with that of an educated intellectual who creates very individual works of poetry, music, sculpture or painting based, still, on skill and knowledge, trained by teachers in specialized institutions, reflecting his own and original ideas which interpret phenomena of the world in a very individual manner.

2.3 "*Artes*"—Fields of Knowledge

Since Late Greek and Roman Antiquity (5th century AD), a system of fields of knowledge was developed in European civilization which became the basis of education until the 18th century. Commensurate with the different human abilities and the principle of division of labor in these highly evolved civilizations, the various fields of knowledge, based on manual and intellectual skills as well, required specialized education or training, and also organization, i.e., they developed into professions.

During the period of Hellenism, from the third century BC to the second century AD, in Alexandria, Egypt at first, and later in the entire Roman Empire, the trend was toward systematization of traditional knowledge. The most prominent learned men compiled voluminous encyclopedias: Marcus Terentius Varro (116-27 BC) described the contents of nine disciplines, a work now lost. It included three chapters on linguistics (at one point called "philology"), four chapters on mathematical disciplines, one chapter on medicine, and one on architecture [Sallmann, 2002]. The most well-known and influential encyclopedia was compiled by Caius Plinius Secundus (23/24-79 AD). His work, Naturalis Historia (including 37 books or chapters), presented the main objects of various scientific disciplines and emphasized useful knowledge of natural objects and phenomena, such as agriculture and horticulture, human and veterinary medicine [Sallmann, 2000].

2.3.1 "*Liberales Artes*" as Basis of Education

These disciplines and architecture as well, considered to be applied sciences at that time, were excluded from the encyclopedia of the Neoplatonist Martianus Capella (c. 470 AD), who defined seven disciplines [Krapinger, 1999] as the basis for the higher education of the youth of high society. Since education and intellectual pursuits not primarily aimed at earning one's living were reserved exclusively to members of high society, i.e., were linked strongly with social rank, they were accessible only to a free citizen or a liberated slave [Christes, 1997; Kuchenbuch, 2001]. Thus these seven disciplines were known as the "liberal arts" (*artes liberales* in Latin). The knowledge comprised in these "arts", disseminated in numerous manuscript copies of Martianus' work, was spread over the whole of Europe by many eminent scholars and founders of new educational institutions. At the beginning of the sixth century, the philosopher Anicius Manlius Severinus Boëthius (c. 480-524) tried to reconcile Neoplatonic ideas with Aristotelian ones. A kind of summary of the contents of the seven fundamental disciplines was composed by the rhetorician Flavius Magnus Aurelius Cassiodorus (c. 490-c. 585), who founded the *Monasterium Vivariense* (near Squillace at the Gulf of Naples in Italy) in 555, in his textbooks on religious and secular knowledge entitled *Institutiones divinarum et De artibus ac disciplinis liberalium litterarum*. Of major importance for spreading the knowledge of the "liberal arts" throughout Europe during the Middle Ages (6th-14th centuries) (for details see [Köhn, 1986]) and the Renaissance (14th-16th centuries)—continued until the 18th century in some regions—was the encyclopedia of the Bishop of Sevilla, Isidorus Hispalensis (c. 560-636) entitled *Etymologiarum sive originum libri XX*. This work, used as a kind of textbook in the early cathedral schools (one of these first European high schools was founded in Chartres, France, c. 1000), in monasteries (for details see [Johanek, 1986]) and in universities founded since the late 12th century, described the principal subjects of learning. The fields of knowledge as the basis of the arrangement of the text were defined.

Based on the works mentioned above, the "seven liberal arts" (*septem artes liberales*) were classified into two groups: The first group, called "the three-way" (*trivium*), included the philological disciplines of grammar, dialectics or logic, and rhetoric. The second group, "the four-way" (*quadrivium*), included the former mathematical disciplines of geometry, arithmetic, astronomy and geography, music or the doctrine of harmonies [Christes, 1997; Bernt, 1980; Hoppe, 1996; Lindgren, 2004]. These "liberal arts" very often figured as motifs for the fine arts, for example as text illustrations of manuscripts and as wall decorations in the reading rooms of libraries.

2.3.2 The "Mechanical Arts"—*Artes Mechanicae*

While fields of knowledge needed for creating useful things by means of manual skills were excluded from the category "liberal arts", they were necessary for sustaining life, and thus were treated, at times, like mathematical and scientific disciplines in manuscripts by learned men. Often they were taught by a "master" or in specialized schools and included in encyclopedias. Following the traditional Aristotelian idea that nature is conquered by engineering, learned men translated the Greek term *mechanikē technē* into *ars mechanica*, thus transforming its meaning in a positive way, pointing out that the implementation of engineering works with the help of manual skills was an intellectual activity as well. They appreciated from the Late Antiquity on (see Cassiodorus in the sixth century and Isidorus in the seventh century), that a human being is able to produce "wonderful" things although all kinds of handicrafts, arts and crafts, applied arts, i.e., "mechanical arts" fashioned by hand, are referred to as "minor" or "easier" arts by some authors until the 12th century. From this time on, the term *artes mechanicae*, including seven different "*artes*", appeared in documents. The learned cleric Hugh of Saint Victor (c. 1130) insisted in his work *Didascalicon de studio legendi* on the statement that an artisan or a craftsman must (or may) have some theoretical knowledge (*ratio*).

Therefore, he interpreted a craft not as *ars* but as *scientia*. He discerned seven categories of crafts necessary for sustaining human life as follows:

1. *lanificium*: weaving, tailoring;
2. *armatura*: masonry, architecture, warfare;
3. *navigatio*: trade on water and land;
4. *agricultura*: agriculture, horticulture, cooking;
5. *venatio*: hunting, food production;
6. *medicina*: medicine and pharmacy;
7. *theatrica*: knights' tournaments and games, theater.

During the Middle Ages, a variety of systems of "mechanical arts" were established by adjusting the selection, nomenclature, and number of these disciplines. By comparing them with philosophy and theology, learned clerics of that time stated that these fields were aimed at providing a livelihood; therefore, they were conceded a lower rank. However, since some practical fields such as the building trade or architecture required the use of geometry and simple mathematical instruments, they were considered by some authors (e.g., Dominicus Gundissalinus, flourished c. 1150) to be applied mathematics, i.e., being equal in value to one of the liberal arts. Several aspects of the mechanical arts or problems of engineering connected with mathematics and physics were treated in separate theoretical writings by medieval authors (e.g., Roger Bacon, Villard de Honnecourt) [Klemm, 1954]. Contrary to certain views, a direct relation between the fields of *artes mechanicae* and the social hierarchy remained until the beginning of the 17th century. Beginning with the period of the Renaissance of Greek tradition introduced by the humanists, their followers valued artistic skill and workmanship, and wrote theoretical works on the mathematical basis for engineering activities (e.g., Leon Battista Alberti, Leonardo da Vinci, Niccolo Tartaglia) and for fine or plastic arts (Leonardo da Vinci, Michelangelo, Albrecht Dürer, and others) [Klemm, 1954; Krafft, 1980].

As supported by a rich body of literature, we affirm that modern sciences have their roots in the systems of "*artes*" as follows: mathematics, astronomy, geography, optics, and acoustics originate in the *quadrivium*, as one sector of the *artes liberales* [Hoppe, 1996], while the *artes mechanicae* comprised the following disciplines: physics, statics and technical fields such as the applied sciences, chemistry, natural history, medicine and others.

2.4 Relationship between the Systems of "*Artes*" and "Arts"

In like manner we find the roots of modern arts in both early education systems. With respect to the Hellenistic system of the seven "liberal arts", its both categories of disciplines must be considered. From the philological group called *trivium* were derived the areas of philosophy, literature, and poetry. The medieval *musica* developed from the doctrine of harmonies into the areas of musical theories connected with the musical practice in different ways. The other mathematical fields of the *quadrivium*, in particular geometry and arithmetic, are reflected by applying geometrical rules, figures, perspective, and symbolic numbers in the fine arts and in the plastic arts.

All kinds of plastic, fine, and performing arts, handicrafts, and architecture as well were practiced from the very early stages of human civilization all over the world and with much intensity during Greek and Roman Antiquity in Europe. With reference to Europe, it is asserted here: after differing judgments on the value of the objects made by humans, they were estimated and evaluated since the 12th century, and were categorized according to the system of knowledge called *artes mechanicae* (mechanical arts). Although the workers in these fields ranked with craftsmen in relatively low social classes until Early Modern Times, their products were in great demand and often made using very precious materials. Since these were natural objects and their derivatives, an artisan had to possess a basic scientific knowledge, e.g., a painter of that time used paint which he made and combined himself from natural products, usually according to a secret recipe.

On the other hand, the collections of artifacts reflect the social standing of the owner in the antique and medieval societies. Persons of the highest social rank possessed sculptures, frescos, books on papyrus, and technical artifacts such as a garden fountain; during Antiquity, these objects were often connected with religious worship and used to decorate a sanctuary or a living space (see such monuments at Pompeii in Italy).

Christian clerics, in the Middle Ages, generally of noble birth or rich feudalistic sovereigns and princes, collected precious artifacts related to their worship, and stored them in treasure vaults. The collections represented the power and wealth of rulers and were significant as a demonstration of a long tradition of nobility and education; moreover, they were a justification for the current high social rank of the owner.

Aside from the demonstration of secular power, the selection and collection of artifacts was influenced by the prevailing cult or religion. Persons and families of the highest social standing decorated their palaces, sometimes also public places, with sculptures, frescos, and technical artifacts; they used them as important gifts in political affairs and kept them in private treasuries. The clerics used the precious artifacts and manuscripts from time to time in religious ceremonies. In ecclesiastical institutions, such as cathedrals and monasteries, some collections of ceremonial artifacts, including rare relics in precious reliquaries, filled the treasure vaults and are still preserved until our time.

2.5 A New Type of Artifact Collection: The Cabinet of Curiosities (14th to 18th Centuries)

By contrast with the artifact collections from Antiquity and the Middle Ages, a turning point in the purpose of collecting valued objects can be observed between the 14th and the 18th centuries. As in earlier times, the most eminent rulers of European countries competed with one another to display the most curious and precious treasures in their residences, in order to demonstrate their power and wealth. For this reason, they supported, in the broad sense, trade with works of art and artifacts. This was carried out systematically by means of a network of agents traveling throughout Europe. These well-trained, often learned persons influenced the selection of objects bought by rich people, such as the French King Charles V in the 14th century, the Emperors of the Habsburg dynasty in Vienna and Prague from the 16th to the 18th centuries, and the Dukes of Bavaria in the 16th and 17th centuries. Since the agents were well-trained, educated, and experienced men, they introduced new ideas into this venture.

The collections of the Early Modern Times reflected not merely power and wealth, but also curiosity about rare objects found in the world [Daston & Park, 1998], and the love of learning and aesthetics by presenting a kind of encyclopedic knowledge. The same intellectual interest served to motivate individual learned citizens and scholars, such as the naturalists Conrad Gessner (1516-1565) in Switzerland, Ulisse Aldrovandi (1522-1605) in Italy, and Carolus Clusius (1526-1609) and Bernhardus Paludanus (1550-1633) in the Netherlands [Hoppe, 1994]. They arranged their collections in separate rooms or even buildings, sometimes called a "museum" and in German "Kunst- und Wunderkammer", in special purpose-built furniture, for their own pleasure and to present their wonderful treasures to their close friends.

Some objects of the most famous encyclopedic collections recall to mind their former splendor and luxury. Many objects of the Imperial "Kunstkammer" in Vienna established in 1558 are found in many different specialty museums today; a reconstruction of the rich cabinet of curiosities is on display in the "Ambras" castle, founded by the Archduke Ferdinand near Innsbruck in Tyrol around 1560. Many of the collections of the Emperor Rudolph II of the Austrian Habsburg dynasty (1552-1612, Emperor from 1576), including paintings, mathematical and astronomical instruments and manuscripts, are kept in the citadel of Prague. The very rich and well-preserved collections established in 1560 by the Elector of Saxony are once again on display in the reconstructed rooms called "Grünes Gewölbe" (the green vault) in Dresden, Germany. The treasures of the Dukes of Bavaria can be admired now in the citadel of Landshut (Bavaria) (for pictures of these famous early cabinets see [Mauriès, 2002]). The collections of the naturalist Ulisse Aldrovandi are kept and displayed in the Museo Civico in Bologna, Italy. Similar smaller collections are featured in many museums all over Europe.

More details on the shape, form, and materials of the objects and of the structure of repositories and rooms are described in original inventories written at the time when the collections were established or shortly after. Since they were compiled mainly by order of the ruler—with the exception of smaller private collections of scholarly men writing the catalogues themselves—they were worked out carefully by learned specialists and experts. Therefore, they are very revealing sources of the history of collecting.

2.6 The System of "*Artes*" as a Principle of Classification of the Cabinets of Curiosities

The scholarly author of the inventory of the ducal "Kunstkammer" in Munich, recently edited according to the manuscript from 1598 [Diemer, 2004], showed clearly that he was familiar with the system of *artes liberales*. Johann Baptist Fickler (1533/34-1610) compiled by order of the Duke the voluminous manuscript, when the rich collection was moved from private rooms in the residence to a new separate building which still exists, but is used now for a different purpose. Coming from Wuerttemberg as a Roman Catholic he was trained from 1551 to 1555 at the University of Ingolstadt, the sole Bavarian University at that time, founded by the Duke in 1472, earning the degree of "*Magister artium et philosophiae*", and at the University of Bologna later on (1564/65), where he obtained the degree "*Doctor utriusque iuris*" in 1565. After working as a secretary to the Archbishop of Salzburg, he became Privy Councillor and private tutor of crown prince Maximilian at the court of the Duke of Bavaria in Munich in 1588. When Maximilian (1573-1651) came to the throne in 1598, Fickler was also a member of his entourage [Diemer, 2004]. In his description of all objects found in the "Kunstkammer", Fickler proceeded through the entire building, room-by-room, and mentioned under others particular objects referring to the *artes*. Among the paintings he found a miniature depicting an act of homage to the Duke Albrecht V (1528-1579) as the patron of arts and civilization in Bavaria, donated by court painter Joris Hoefnagel (1542-1600) in 1579 [Diemer, 2004; Diemer et al, 2008]. It shows at the bottom a view of Landshut and at the top one of Munich; in the central area a green tree looms up in front of a garden in the background. Besides the tree stand two women interpreted by Fickler as "*Musæ seu Artes liberales*" and identified with help of their attributes as "*geometria*" and "*musica*". Moreover, several verses of poems by the Roman poet Virgil are quoted. The ornaments in this painting in the Renaissance style (stored in the "Staatliche Museen, Preussischer Kulturbesitz,

Kupferstichkabinett," Berlin) demonstrate that not only Fickler but also the earlier painter were well-trained in the classical antique tradition.

After the middle of the 16th century, another learned man was employed by the Duke of Bavaria in Munich. Samuel Quicche(l)berg (1529-1567) was born in Antwerp in Belgium. He worked first with the wealthy Fugger family, in particular with Johann Jakob Fugger in Augsburg, and was appointed in 1559 by Albrecht V, the Duke of Bavaria, probably as a physician in ordinary, after working for the Duke as a collecting agent since 1553. Quicche(l)berg was active as an agent for rare artifacts with international connections (especially in western and southern Europe) and may have visited several private cabinets of curiosities in his native country, in Italy, and in Basel, Switzerland, where he was enrolled at the university. After the ducal "Kunstkammer" was established and transferred from the residence to a new special building erected between 1563 and 1567, he wrote a kind of a catalogue with a classification of the main objects to be included in an ideal cabinet of curiosities, published in Munich in 1565 [Quiccheberg, 1565] and entitled *Inscriptiones vel Tituli Theatri Amplissimi, Complectentis rerum vniuersitatis singulas materias et imagines eximias, […]: Promptuarium artificiosarum miraculosarumque rerum, ac omnis rari thesauri et pretiosæ supellectilis, structuræ atque picturæ, quæ hic simul in theatro conquiri consuluntur, ut eorum frequenti inspectione tractationéque, singularis aliqua rerum cognitio et prudentia admiranda, citò, facilè ac tutò comparari possit* (Inscriptions or Titles of the Most Extensive Theater, Comprising the single matters of the universe of things and extraordinary images […]. Archives of manmade and wonderful things, and also of the structures and paintings of the entire curious thesaurus and the precious equipment, which are recommended here for collecting in the theater simultaneously, in order that a certain singular knowledge of things and an admirable prudence by means of their frequent inspection and treatment can be obtained quickly, easily, and certainly [English translation by the author]) (A German translation and comments, mentioning in passing the traditional ideas of *artes* in this work, is found in [Roth, 2000].)

The very title of this work not only gave rise to the abundance of different objects representing the most curious and rare things in the world, but also mentioned explicitly the most important aim of such a collection: a cabinet of curiosities was meant to instruct; the eager visitor should be given new knowledge about various things by being allowed to inspect the objects frequently. When examining the detailed lists of objects as found in Quiccheberg's project on an ideal all embracing ("universal") museum with the objects arranged according to the systems of liberal and mechanical arts, we discover clear parallels to both systems in the five classes with their sections and further chapters. These canons of scholarly disciplines are reflected not only in the many statements of the text, it also refers to other well-known traditional classification modes such as the seven planetary gods with their numerous symbolic meanings [Roth, 2000], the four elements, four seasons and phases of life. It required a certain level of scholarliness and special training to identify the symbolic significance of many of these objects. The arrangement of the major objects into the five "classes" of Quiccheberg's catalogue as compared to the systems of *artes liberales* (*a.l.*) and *artes mechanicae* (*a.m.*) may be analyzed as follows:

I.	Paintings on the genealogy and dominions of the current ruler		– *a.m.*
II.	Wood, stone and metal sculptures		– *a.m.*
III.	*Naturalia*: animals, humans, plants, minerals, metals		– *a.m.*
		– natural and manmade objects in their settings	
IV.	Instruments: musical, mathematical, astronomical		– *a.l.*
	"	: tools for writing and painting	– *a.l.*
	"	: mechanical, medical, agricultural; ships	– *a.m.*
	"	: arms, textiles, games	– *a.m.*
V.	Paintings, engravings, gobelins and tapestries		– *a.m.*

Similar categories, and some of the objects, can be found in many other cabinets still existing, or documented in catalogues, clearly indicating the relationship between the cabinets of curiosities and scholarship under the influence of humanistic education during the Renaissance period.

2.7 The "*Artes*" Concept Approved and Acquired by Artists and Craftsmen

Innumerable artists and craftsmen created individual objects and displays for cabinets of curiosities, established from the 15th and 16th centuries by wealthy personages. The rooms of the cabinets were decorated with allegoric figures from the artes, the seven planets, the four elements, four seasons, four phases of life, five senses; all these themes were represented by many symbolic figures and objects. Both the collectors and the artists were familiar with these systems of learned knowledge. The tendency to possess these cabinets of curiosities spread throughout all of Europe via persons of noble birth, including many clergymen, and rich citizens of certain smaller republics. It was for this type of client that several painters in Antwerp created the motif, "painting of a gallery"as from the early 17th century. Such paintings were done by Hans Jordaens III (who flourished c. 1630), the original of which can be seen in the Museum of Fine Arts in Vienna; by Jan Breughel the elder, the so-called "Sammet Breughel" (the velvet-Breughel) (1568-1625), working at times with P. P. Rubens, whose masterpieces are kept now in the Museum Prado in Madrid; and by others. These large paintings showed the interior of a big cabinet with paintings on the walls and floor, sculptures, globes, books, shells of different shapes, medals and coins, gems, jewels, glasses found on furniture. Many visitors are looking at the abundant objects of art and the crafts, both artificial and natural. Often an open window or a door in the background gives a view of a garden toward a landscape that signifies nature in its cultivated state.

When the passion for collecting curiosities and exotic items peaked around the middle of the 17th century, some known and sometimes wealthy artists established private cabinets of curiosities serving as models. One of the most eminent collectors was Rembrandt Harmenszoon van Rijn (1606-1669). Not many objects survived from his original collection of contemporary paintings, drawings, and etchings, as he was forced to sell them for lack of money later in life. He had one of his homes in the center of Amsterdam equipped with one room full of display cabinets for his collection of curiosities including arms, armors, metallic vessels, minerals, shells, small sculptures of human and animal figures, and of individual human organs for the study of anatomy. The reconstruction of Rembrandt's former study in his home, now a museum in Amsterdam, recalls the intensity which he devoted to maintain and add to his collection, although the reconstruction and repeated restorations of the entire house altered much its original state [Vries de, 2007].[1] Many of the utensils found in his paintings are intended to signify certain attributes of the relevant person. One of the smaller objects, a shell of approximately fifteen centimeters in length, with attractive shape and patterns, called in Latin *conus marmoreus* (marbled cone), was immortalized in an etching by Rembrandt in 1650 (Fig. 2.1) [Biörklund & Barnard, 1968, no. BB 50-E].

Fig. 2.1. Rembrandt: Etching of the shell *conus marmoreus*, signed 1650 [Biörklund & Barnard, 1968, no. BB 50-E].

2.8 Scientific Motifs in the Fine Arts

After sculptures often idealized and displayed eminent scientists and philosophers for centuries, individual portraits were created beginning in the period of Renaissance. The objects which signified certain attributes, such as an armillary sphere, a compass, a book, a retort or a flower, immediately revealed the profession of the person in the portrait, but in a superficial way. Some artists emphasized which type of a scientist this

[1] For a picture of it, see p. 191, Fig. 14 in [Vries de, 2007].

was. A well-known example is a work by Rembrandt dated around 1652. In this etching entitled "Faust" or the "alchemist" [Biörklund & Barnard, 1968, p. 116, no. BB 52-4] he delineated a scientist, having searched for a mystery in nature, with a vision represented by an anagram.

Extraordinary scientific discoveries documented by painters demonstrate that these were very experienced people who participated in truly current learned discussions at the time. The National Museum of Fine Arts at Vienna has an oil painting by Giorgio da Castelfranco, known as Giorgione (1478-1510 Venice), entitled "Three Philosophers". After a number of intensive but inconclusive interpretations of its meaning published by philosophers and art historians [Hirdt, 2002], the enigma was solved around five years ago. The three figures—the two oldest standing and discussing, and the youngest sitting and observing a natural phenomenon, are seen on a step-like rock on the edge of a wood with a view opening on a sunny landscape in the background. The most interesting figure is the old man with a long grey beard, who holds in his two hands an astronomical document and a compass. The careful analysis of this document containing astronomical symbols (e.g., of a crescent moon), reveal to us certain Latin and Greek characters and a number of Latin ciphers. In the right upper corner the word "Jovis" is written and just below this word four dots appear. Partly covered by the shadow of the fingers, the following date is readable: 15/4/[150]5. The Greek word in the middle is "selēnē", the moon; it refers to the title of a work of the most famous astronomers of Greek Antiquity, Aristarchus of Samos (c. 310-c. 230 BC), who assumed that the solar system was heliocentric. The man standing in the middle of the group and wearing a turban is thought to be an oriental scholar discussing with the Greek, since the first translator of the work of Aristarchus was a Moslem scholar [Stahl, 1970]. The third person, the young man seated and observing the natural surroundings, seems to embody early modern empirical science. This recent identification of the three personalities on the painting in a historical way is clear and straightforward.

Based on this recent interpretation we assert that the significance of the astronomical notes presented by the figure of Aristarchus is really remarkable, and means the following: The four satellites or moons of the planet Jupiter (depicted as the four dots below the word "Jovis" in the right upper corner of the white leaf) were observed in Venice, Italy (where Giorgione lived at that time) by scholars in April 1505 [Keim, 2009], who ought to have realized the importance of this discovery. It showed the phenomenon of a high planetary sphere in the heavens, observed in the earthly region of the universe and known, until that time, to belong only to that region. This phenomenon was strictly contrary to the traditional Aristotelian doctrine which contended that there was maintained a strong separation between the "translunar" and the "sublunar" part of the universe [Hoppe, 1998]. It appears that these Venetian scholars had discussed this discovery in their own circle, where Giorgione may have learned about it. Moreover, the recent interpretation of the painting may be sensational as it relates to the historiography of astronomy. Scholars in this field have until the present taught that no less a person than Galileo Galilei (1564-1642), when he was a professor at the University of Padua, was the first to discover the four satellites of Jupiter (in 1609, published in 1610) with the aid of his telescope "prepared" by himself, using Venetian glass lenses [Drake, 1978]. The painting of Giorgione reveals that this astronomical discovery, which had a supporting revolutionary effect on Aristotelian science, occurred in Venice a hundred years earlier than recorded in historical textbooks to the present day (see, e.g., [Dreyer, 1953]). Since the Venetian scholars may have recognized the revolutionary significance of their observation, they did not publish it immediately, perhaps also for lack of a scientific journal, something which was not "invented" until around 1500. Thus this revelation in the form of a small note on a large painting was and remains an important documented witness to a major scientific discovery.

Another kind of astronomical observation, made with the aid of the newly invented and immediately improved telescope around 1600, was documented very early by the German painter Adam Elsheimer (1578-1610), living in Rome, Italy, at that time. He painted the old biblical motif of "The Flight into Egypt" in an unusual manner, i.e., the scene occurred during the night. Jesus' mother Mary held the holy child in her arms while sitting on a mule, with Joseph walking and lighting the way with a lamp. Nearly half of the surface of the painting was taken up by a clear night sky. On the horizon the full moon is reflected in the water of a lake in one corner. Stars dotted the sky, arranged in such a way as to allow some of the constellations to be identified. Starting in one corner, a ribbon of light shone through the sky. It was composed of innumerable tiny bright dots, the stars. Most likely inspired by contemporary astronomers, the painter depicted the exact configuration of the Milky Way as known in 1609 (the original painting is in "Alte Pinakothek" in Munich). Although this phenomenon had been known since the oldest advanced civilizations, e.g., in Egypt, it could not be observed precisely enough without the use of a telescope. Therefore, the astronomers assumed the Milky Way to be of a nebulous character [Dreyer, 1953]. This accurate, and aesthetically very impressive portrayal by Elsheimer soon spread throughout Italy, the Netherlands and Germany via several copies in color and in black and white [Baumstark, 2005]. The copy consisted of a copperplate-engraving and etching (1613), the masterpiece of Elsheimer's student, Hendrik Goudt (1583-1648), and in particular highlighted the ribbon of light representing the Milky Way. This early copy became itself a prototype for contemporary painters [Baumstark, 2005] (see also [Renger, 1982]).

An important and strong relation between the natural sciences and the fine arts developed during the period of voyages of discovery originating in Europe in the late 15th century. The traveling naturalists—not all were gifted and trained enough to do it themselves—became dependent on a good draftsman and painter for documenting their numerous observations in the foreign countries overseas, relating to geography, landscapes, the human inhabitants and their way of daily life, flora, fauna, minerals, natural products, including how they were processed and how they were used. The artists accompanying the travelers discovered new motifs and became aware of previously unknown shapes, colors, light, shadow and other effects. This kind of collaboration between scientists and artists, fruitful for both sides, continued until the end of the 19th century and was only partially replaced by the invention of photography. From the numerous results of that kind of collaboration I will describe here only one of the early examples, when European explorers discovered unknown parts of the earth, far from European civilization, such as the primeval forests of South and Middle America, which some of them referred to as "paradises lost".

The Dutch painter Frans Janszoon Post (1612-1680) traveled to Brazil in 1636 together with the embassy of Prince Johan Maurits of Nassau-Siegen (1604-1679), who was the governor-general of a Dutch colony in Brazil (reigning there from 1637 to 1644) and who acted on behalf of the Dutch West Indian Company. During his stay and possibly until the beginning of 1644, he drew and painted in very precise detail many typical subjects as perceived by a European painter for the first time, such as actual locations, landscapes with tropical vegetation, native inhabitants, and their working places (Fig. 2.2). Sometimes he only sketched the objects of study which he elaborated into paintings after returning to Haarlem, the Netherlands. His works have spread widely to museums all over the world now and are much esteemed by reason of their authenticity and high artistic quality [Whitehead & Boeseman, 1989; Corrêa do Lago, 2007]. Another Dutch painter, Albert Eckhout (1610-1666) of Groningen, traveling together with F. Post from 1636 to 1644 to Brazil, often did still-life paintings of single objects, such as fruits, animals, and even exotic-looking human beings. Moreover, the ethnographic documents he compiled were his greatest contribution to posterity, as they included the very earliest illustrations of those themes [Whitehead & Boeseman, 1989], and their subjects, namely the native people and their living environments in the 17th century, which have now entirely disappeared.

Fig. 2.2. Frans J. Post: Ink drawing showing a Brazilian Sugar Mill; original in the "Musée des Beaux-Arts", Brussels (after Plate 101a in [Whitehead & Boeseman, 1989]).

The view of such native people in their daily-life, natural environment presented by these artists was that of a foreign observer, coming from a vastly different social world; it was not really objective, but very much influenced by a colonial attitude. This view was identical with that of scientists at that time. Both scientists and artists were hired by a European prince and shared the same purpose: to observe and explore these far-away territories and their inhabitants in order to extend the own knowledge and the secular power of the sovereign who strove to derive maximum benefit and profit from these ventures.

The concept of scientific achievements by an artist at home, within his own social environment, should be expressed in another manner. During the Dutch Golden Age the traveling naturalists brought home not only drawings and other kind of pictures of natural objects, but also real fruits, seeds and young live plants. They sowed and cultivated them in Botanical Gardens (refer to the leading role of Carolus Clusius, the botanist at Leiden University after 1593) and thus developed the horticulture of the country. The multicolor and manifold results of cultivation were acclaimed in the numerous paintings of flower bouguets by the most talented artists (see for example, the still-life paintings by Ambrosius Bosschaert, 1573-1621). At the same time these Dutchmen developed trade with flowers and plant products which grew continually in economic importance. Rare bulbous plants, in particular the tulip brought with the help of the Leiden botanist Clusius from Turkey to Europe and successfully cultivated in the West, allowed the Dutch to profit significantly from the 1620s to the 1630s, as they created a market for durable tulip bulbs. When the bulb and flower trade peaked in the wake of economic speculation, a single buyer offered several thousand florins for one special variety of a tulip, i.e., more than a skilled craftsman earned in one year (c. 150-300 florins). Many wealthy citizens (especially businessmen from Haarlem) soon became involved in extreme stock-market dealings. However, prizes for purchase contracts burst like a bubble in February 1637 [Dash, 1999]. Although relatively few people lost money, a broad discussion on the real value of such a transient object as a flower bulb ensued among the Dutch citizens, very much influenced by the Christian view that tulip mania or "tulipomania" was a perversion of the moral order.

Those events also inspired some painters to participate in the discussion via their professional work. A very impressive painting directly exposing the contemporary perversion of the tulip trade was done by Jan Breughel the younger (1601-1678) around 1637. His "Allegory of Tulipomania", original in the Frans-Hals-Museum at Haarlem (the Netherlands), was probably part of a series of paintings on moralistic themes known as "Allegories of the Topsy-Turvy World", created in the early 1640s [Ertz, 1984]. In the Haarlem painting we see several scenes in the foreground between steps to a balcony of a house at the left and a terrace at the right. The view in the background opens to a short avenue in the middle, ending with a gate in Greek classical style and on both sides meadows extend, bordered by woods. All of the figures, including two horse-riders in the background, in scenes showing the various phases of selling and buying tulip bulbs, are not human beings, but monkeys with long tails. The phases of the tulip trade having a contrary outcome are depicted on the opposite sides of the painting, one group on the left and the other on the right. The positive stages of the trade, from the inspection of a tulip bed by a potential buyer, through weighing the bulbs on a scale, calculating the price, drawing up a contract, counting the profits and then enjoying a banquet under a picture of a tulip flower after a successful conclusion of the business, are shown as a contrast to the negative outcome of a failed deal, with the marching off and punishment of a defaulting purchaser.

Although the criticism of this painting and others on the same theme pleaded by several contemporary painters and also many pamphlets concerning the subject of tulip mania were primarily targeting the economic abuse of the love of flowers, they were also directed at its scientific basis. The enthusiastic craftiness of the bulb breeders and the propaganda for their products made it possible for the trade to flourish. The attention focused on these events might serve to admonish botanical scientists to seriously consider the possible effects of their work on contemporary society.

2.9 Turning Points in Arts and Sciences in Modern Times

The strong advancement in the sciences based on empirical evidence and changes in epistemology was followed by technical inventions leading to a technological and industrial revolution in Europe from the 18th to the 19th centuries. It affected human living conditions and changed social structures and political cultures little by little until the middle of the 20th century. The different types of revolutions had rather a major impact on artists, who grappled with the problems by means of what they created, striving to develop an individual manner of expression by interpreting, through their art, the events occurring around them [Goodman, 1984].

2.9.1 *Turning Point Following the Enlightenment Philosophy*

While the traditional *artes* canon remained the basis for university education in Europe, and although many of the finer aspects of these disciplines had been altered [Hoppe, 1994], several individual philosophers and scientists developed a new epistemology of science and the philosophy of nature by criticizing in particular the Aristotelian philosophy of science. Prominent thinkers of the 17th and 18th centuries, such as R. Descartes, F. Bacon, the German J. Jungius, I. Newton, G. W. Leibniz, D. Diderot and J. le Rond d'Alembert, to recall only the most important among them, established the groundwork for the so-called "new science", including empirical and experimental results investigated using quantitative and statistical methods, and interpreted ultimately by mathematical means. During the scientific revolution from the 17th century onward, the scientific and mathematical disciplines developed their own methodology based on hypotheses and probability [Daston & Park, 2006].

An iconical turn had taken place since the Renaissance period, although the new ideals of aesthetics gained ground only slowly. The traditional religious motifs were left aside as painters pursued new subjects, and discovered individual portrait painting. They began to arrange persons and single objects differently in a picture, making use of new mathematical methods, in particular the new understanding of perspective (see L. B. Alberti in the 15th century, A. Dürer from the 15th to the 16th centuries, and their followers). At first the versatile Italian Renaissance artists such as Alberti and Leonardo da Vinci, at the end of the 15th century, underlined the high rank of fine arts, in particular of painting, by comparing it in their theoretical treatises with poetry. They and their followers praised the ingenuous imagination of a painter whose professional standing became similar to that of a scholarly man. This tendency increased in the French "classicism" from the 17th century [Ullrich, 2005]. We find certain characteristic tendencies and features in the European fine arts. The artists tended to develop an individualistic style and handled current social problems in their art. The landscape and still-life painting (using natural objects as motifs), in particular in the Netherlands and France during the 17th century, became widespread in the 18th century. Persons representing a traditional mythological or religious theme were placed in an open landscape, though it was often idealized and kept to the side. Artists, often in collaboration with scientists, illustrating scientific texts or making rough sketches of newly discovered natural objects, as well as scientists on their own continued to observe closely natural phenomena (see the development of natural history in the work of G. L. Leclerc Comte de Buffon and the illustrators of his publications) and experiments in and with the natural environments of plants (J. G. Koelreuter, J. Priestley, A. v. Humboldt) [Hoppe, 1997; Hoppe, 2000] and animals (R. A. F. de Réaumur, A. Trembley, C. F. Wolff). Philosophers participated in this during that "Age of the Cult of Nature" [Klonk, 2003]. In addition, the arts began to adopt subjects such as the daily life of humans of all social classes as well as contemporary social and even historical themes. Not only the motifs but also the styles changed in the search for new aesthetical manners of expression. Artists tried to interpret subjects and events in their contemporary world in an individualistic way from the 19th to the 20th centuries.

After the development of sciences and the technical applications of these results, the first technical and industrial revolution spread across the European continent and to North America from the 18th to the 19th centuries [Friedel, 2007]. Soon artists found new topics, documented technical inventions and interpreted their effects on human society. The invention of the hot air balloon by the brothers Jacques Etienne (1745-

1799) and Joseph-Michel Montgolfier (1740-1810), who released a hot air balloon into the air at Annonay, France in June 1783 for the first time, was a public event witnessed by numerous spectators and covered in many newspapers. After some technical improvements, a balloon carrying two or three men became a useful means of transport, even during the combat of the French Revolution. Thus it very quickly captured the attention of many people all over Europe. One eminent painter of that period depicted the extraordinary view from up in a balloon of the landscape and the people watching it from the ground. The Spanish painter Francisco José de Goya y Lucientes (1746-1828), who portrayed many actual political and social subjects [Gudiol, 1971; Carrete Parrondo, 2007], created a drawing in black chalk on paper entitled "The Montgolfier Brothers" (the original, 27 × 38 cm, is found in the "Hamburger Kunsthalle" at Hamburg, Germany). Two thirds of the canvas is taken up by the sky and the enormous balloon in the foreground, while the people on the ground look small and insignificant, and a mountain range on the horizon is a mere strip. Between 1800 and 1808 he painted in oil on canvas another "Ascent of the Montgolfier" or of the "Balloon" (the original, measuring 83 × 103 cm, is now in the "Musée des Beaux Arts" at Agen, France). In this work he presented the view of a hilly landscape as seen from a balloon with crowds of onlookers on a plain in the middle of the fore-ground. The ascent of a huge captive balloon with one of the passengers wearing a silk hat impressed the German painter, Adolph Menzel (see below) in the middle of the 19th century. He was much affected by the way that the big balloon diminished in size with increasing distance [Wirth, 1965].[2]

This painter Adolph Menzel (1815-1905), living in Berlin, was widely known for his naturalistic works showing the daily life of his middle-class family on one hand, and scenes from the Prussian Royal Court of Frederick the Great, and later on the Imperial Court, as well as important events from high society, on the other [Hütt, 1981]. He became much interested in technical inventions and the social changes in the wake of industrialization. He painted a landscape near Berlin showing, in

---

[2] See figure on p. 152 in [Wirth, 1965].

the center a long curved track of the first railway between Berlin and Potsdam, a train pulled by a locomotive with a steam engine emitting a long trail of grey smoke (see the "Berlin-Potsdamer Bahn" painted in 1847; original at "Staatliche Museen, Preussischer Kulturbesitz, Nationalgalerie," Berlin). The smoke was carried by the wind across the plain and this bird's-eye view provided a most unusual picture of the landscape. Menzel carefully documented both technical details and those relating to human behavior, which he studied first-hand while touring Prussian factories and observing the installations at close range (Fig. 2.3). The result was one of the first paintings of a modern factory hall from 1872 to 1875 based on an iron rolling mill in Königshütte, Upper Silesia. The painting, now known as "The Iron Rolling Mill (Modern Cyclops)" (in German: "Das Eisenwalzwerk, Moderne Cyclopen"; size: 158 x 254 cm) was in private possession for one year and was then bought by the recently opened National Gallery of Berlin in 1876 [Hauser, 2007]. In this painting Menzel depicted the hard labor of workers in the iron industry of the 19th century, in enormous halls at high temperatures from an open fire-place and dominated by big machines. This painter was skilled at conveying the atmosphere of the working place and the external conditions of labor in a modern industrial factory. Such a painting is also intended to prompt a critical discussion of the negative and dangerous aspects of progressive technology, concerning both the physical and health dangers of the workers.

The discussion of the impacts of technology and industrialization was continued by many artists during the 20th century using various styles. Under the influence of deep social and political changes, artists became more sensitive to social themes of human life. In particular, after World War I they depicted their experiences, sometimes "themselves shell-shocked", by creating individual works of art. Along with the subsequent dominance of abstract and non-figurative presentations, there arose a tendency toward individualistic and unusual arrangements of shapes, materials, and color combinations. New artistic techniques dealt with both the manner of presentation and of artistic statement. Shapes and colors took on an autonomous quality. The products of the fine and plastic arts are so multi-faceted that a blanket judgment does not offer any clarification. We will mention here only a few, but classic, examples of works showing the effects of technology on human life.

Fig. 2.3. Adolph Menzel, *In the Royal Iron-Foundry at Berlin*, 1874, lead-pencil drawing on paper; original at "Staatliche Museen, Preussischer Kulturbesitz, Nationalgalerie", Berlin (after [Wirth, 1965, p. 132]).

The way of life in a modern city with tall buildings and small streets in the center, or crisscrossed by highways and many cars, and a trolley network and all that goes with it, may cause an individual to feel lost on his own, even in crowds of other human beings. Similar subjects and situations were depicted by the following painters: the young, highly gifted Maurice Utrillo (1883-1955) and Albert Marquet (1875-1947), both of whom painted the streets of Paris in a late impressionistic style; the Italian cubist Gino Severini (1883-1966) and the French painter Robert Delaunay (1885-1941), with his cubistic compositions; the German emigrants Max Beckmann (1884-1950) and George Grosz

(1893-1959), the latter in particular in his expressionistic paintings from 1916-1917 entitled "Big City" ("Großstadt", in the "Collection Thyssen-Bornemisza", Madrid) and "Metropolis" (Museum of Modern Art, New York) showing the former "Central-Hotel" in Berlin with the street crossing at "Friedrichstrasse", where a crowd of human beings hurried between a trolley line and a hearse, the entire scene bathed in a red light. Similarly, the daily labor of the majority of workers, dramatically transformed by technical inventions and the subsequent organization of industrial production, was clearly portrayed in many works of art in the first half of the 20th century.

2.9.2 *Turning Point Based on Technical and Industrial Development, and Accompanied by Social and Political Changes*

In the 1920s and early 1930s, a circle of "progressive" artists was founded in Cologne, Germany, persons who aimed to demonstrate in their art, the universal aspects of industrialization as opposed to the individual ones. They tried to create a "socialistic art" in the sense of the "class-struggle", given expression in popular media, such as newspapers and even signboards. They focused not only on social themes, but also on the art of painting critically, searching for an original manner of presentation [Hemken, 2008]. One member of this group, Franz Wilhelm Seiwert (1894-1933), created paintings and black-and-white woodcuts showing workers at machines in a factory room [Roth, 2008]. The shapes of both, the humans and the machines, are reduced to geometric figures: The main parts of the machines and of the human bodies look like squares connected by curved tubes; instead of faces, the cylindrical heads of the humans and the machines have numbers. The "human" heads are connected by a large appendage from a machine, resembling a swivel arm with a center of motion shaped as a circle, whose diameter is greater than that of the heads. This signifies that the factory machines dominate the workers, who must function like parts of a machine ("Die Fabrik", a painting from 1923, now disappeared).

A much more positive view of the modern world of labor with machines and technical products was offered by the French artist

Fernand Léger (1881-1955) [Bauqier, 2003]. He combined a machine, for example a bicycle, and a scaffold, not only with humans but also with natural objects, such as flowers or fruits and branches of a tree and clouds (see the series of works entitled "Construction Workers" from the 1950s). Although the workers are strongly connected with and even a part of their working place and its potential dangers, their environment becomes broader. If they clamber up the scaffold, they can climb higher than a tree and can reach, with the help of the scaffold, a domain nearer to the sky and the clouds. The world of workers with technical gear and tools looks cheerful and is radiantly, colorfully illuminated.

While many painters, F. W. Seiwert, G. Grosz, F. Léger and others, very much simplified and stylized the shapes of objects to look similar to geometrical figures in their works, these shapes nevertheless still represent the recognizable figures of human beings or natural and technical objects. An enduring turning point in the modern arts occurred around 1910, when Wassily Kandinsky (1866-1944) created his first abstract paintings and published his book "Concerning the Spiritual in Art" ("Über das Geistige in der Kunst") [Kandinsky, 1912], followed by his theoretical analysis "Point and Line to Plane" ("Punkt und Linie zur Fläche") [Kandinsky, 1926] and many theoretical articles in the 1920s [Friedel, 2008]. By discussing earlier theorists of art, Alberti, Leonardo da Vinci, and others, he developed his ideas and created pictures using elementary geometrical shapes, points, straight or curved lines and clearly colored planes, delineated by different shapes (Fig. 2.4). Kandinsky's ideas became extremely fruitful during his professorship at the "Bauhaus" art school at Weimar and Dessau, Germany, from 1922 to 1933, where his colleagues Paul Klee (1879-1940), Oskar Schlemmer (1888-1943), Willi Baumeister (1889-1955), and others worked in a similar manner. Until the end of their lives, Kandinsky and Klee in particular created innumerable new abstract shapes in novel color combinations.

Similar ideas in the abstract style of painting and plastic art were developed and carried out by Piet Mondrian (originally Pieter Cornelis Mondriaan, 1872-1944) and Theo van Doesburg (originally Christiaan Emil Marie Küpper, 1883-1931), both from the Netherlands. Together they founded the group of artists called "De Stijl" in 1917. Van Doesburg was a teacher at the "Bauhaus" from 1921 to 1923, wrote on

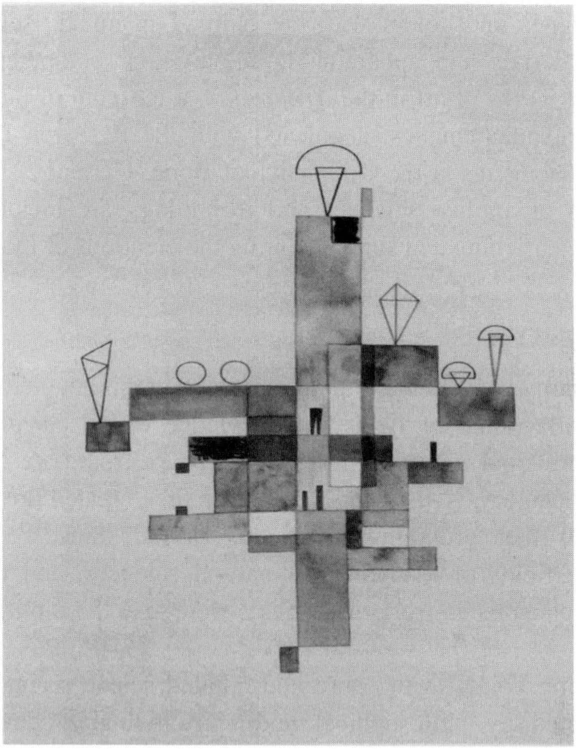

Fig. 2.4. Wassily Kandinsky, *Light Cubes* (*Leichte Kuben*), 1932, watercolor on bluish green paper, size: 46,9 × 28,1 cm; original in the collection of Harry Lewis Winston, Birmingham, Michigan, USA (after Tab. 71, p. 103 in [Johnson, 1964]).

the topic of the new aesthetics ("Von der neuen Ästhetik zur materiellen Verwirklichung") in 1911, and published a "Manifesto of the Concrete Art" ("Manifest der Konkreten Kunst") in 1930. These at the same time artistic and intellectual movements became very influential and were further developed in the works of the Hungarian artist László Moholy-Nagi (1895-1946), teaching at the "Bauhaus" 1923-1928, and of the Swiss Max Bill (1908-1994), studying at the same school in 1927/28. The latter artist developed and pursued "Concrete Art" with many young artists, architects, and designers [Gomringer & Hofmann, 1991]. Their followers in the present day use not only natural but also modern artificial materials and objects; and some of these artists use electronic techniques, or become inspired by the possibilities to create moving and changing shapes and colors with the help of manmade scientific and technical materials and apparatuses, in particular computers (see the exhibition of works of art in the *Arithmeum*, a museum of the history of reckoning and mathematics, connected with the Research Institute for Discrete Mathematics at the University of Bonn, Germany). Nowadays, the products of applied sciences and technology are included among works of art, enriching and supplementing the creations of modern artists.

2.10 Summary

Here is a summary of the major points covered in this chapter.

1. A wide range of fields of knowledge called "*artes*" in Latin formed the principal education program in Europe from Late Antiquity to the mid-18th century. Artists and the objects they created drew from the principles of that program during this period. Since "knowledge" included the fields of mathematics, natural sciences, and technology, objects and phenomena from these fields were also the motifs of artistic works.

2. Turning points in the pure and applied sciences and in the arts were prompted by philosophical and epistemological developments, giving rise to changes in the living conditions of human beings in Modern Times.

3. In line with the education system of the day, during the course of the history of European civilization, scientists and artists cultivated common attitudes toward natural phenomena:

- During the Hellenistic and Renaissance periods, the purpose of the sciences was to discover the essentials of nature; in particular, applied sciences and the fine and plastic arts as well strove to imitate nature (*imitatio naturae*, in Latin).
- In Modern Times, sciences and arts, each using its own special methods, seek to interpret nature (*interpretatio naturae*, in Latin) (compare also with [Goodman, 1984]).
- After a long period of criticizing technological inventions—though they were based on physical laws, strictly speaking—and their impacts on human life and society, certain artists in our days

represent positive effects of technology. Sometimes, they use chemical substances and physical apparatuses in their works of art.

References

Baumstark, R. [2005] (ed.) *Von neuen Sternen. Adam Elsheimers Flucht nach Ägypten* (Pinakothek-Dumont, München).

Bauquier, G. [2003] *Fernand Léger. Catalogue raisonné de l'oeuvre peint, 1949-1951*, Vol. **8** (Adrien Maeght, Paris).

Bernt, G. [1980] "Artes liberales I-II," in *Lexikon des Mittelalters*, Vol. **1**, ed. Avella-Widhalm, G. et al (Artemis, Zürich) col. 1058-1061.

Biörklund, G. & Barnard, O. H. [1968] *Rembrandt's Etchings True and False. A Summary Catalogue*, 2nd edition (George Biörklund, Stockholm).

Burland, C. A. [1988] *The Arts of the Alchemists* (Macmillan, New York).

Carrete Parrondo, J. [2007] (ed.) *Francisco José de Goya y Lucientes. Estampas: grabado y litografia* (Electa, Barcelona).

Christes, J. [1997] "Artes liberales," in *Der Neue Pauly. Enzyklopädie der Antike, Altertum*, Vol. **2**, eds. Cancik, H. & Schneider, H. (J. B. Metzler, Stuttgart, Weimar) col. 62-64.

Corrêa do Lago, P. [2007] *Frans Post* (5 Continents Edition, Milano).

Dash, M. [1999] *Tulipomania. The Story of the World's Most Coveted Flower and the Extraordinary Passions it Aroused* (Crown, New York).

Daston, L. & Park, K. [1998] *Wonders and the Order of Nature, 1150-1750* (Zone Books, New York).

Daston, L. & Park, K. [2006] (ed.) *Early Modern Science* (*The Cambridge History of Science*, **3**) (Cambridge University Press, Cambridge)

Diemer, D. et al [2008] *Die Münchner Kunstkammer*, Vol. **1-3** (*Bayerische Akademie der Wissenschaften, Philosophisch-Historische Klasse, Abhandlungen; Neue Folge*, **129**) (Bayerische Akademie der Wissenschaften, in Kommission bei C. H. Beck, München).

Diemer, P. [2004] (ed.) *Johann Baptist Fickler. Das Inventar der Münchner herzoglichen Kunstkammer von 1598. Editionsband, Transkription der Inventarhandschrift cgm 2133* (*Bayerische Akademie der Wissenschaften, Philosophisch-Historische Klasse, Abhandlungen; Neue Folge*, **125**) (Bayerische Akademie der Wissenschaften, in Kommission bei C. H. Beck, München).

Drake, S. [1978] *Galileo at Work. His Scientific Biography* (University of Chicago Press, Chicago).

Dreyer, J. L. E. [1953] *A History of Astronomy from Thales to Kepler*, 2nd edition (Dover, New York).

Ertz, K. [1984] *Jan Breughel der Jüngere (1601-1678). Die Gemälde* (*Flämische Maler im Umkreis der großen Meister*, **1**) (Luca, Freren).

Friedel, H. [2008] (ed.) *Kandinsky: absolut, abstrakt* (Prestel, München).

Friedel, R. [2007] *A Culture of Improvement. Technology and the Western Millenium* (MIT Press, Cambridge, MA).

Georges, K. E. [1879] *Ausführliches Lateinisch-Deutsches Handwörterbuch*, Vol. **1** (Halm'sche Verlags-Buchhandlung, Leipzig).

Gomringer, E. & Hofmann, S. [1991] *Museum für Konkrete Kunst Ingolstadt* (Georg-Westermann, Braunschweig).

Goodman, N. [1984] *Weisen der Welterzeugung* (German translation by Looser, M. of »Ways of Worldmaking«, Hackett Publishing Company, Indianapolis, Cambridge, 1978) (Suhrkamp, Frankfurt am Main).

Grote, A. [1994] "Vorrede—Das Objekt als Symbol," in *Macrocosmos in Microcosmo. Die Welt in der Stube. Zur Geschichte des Sammelns, 1450 bis 1800* (*Berliner Schriften zur Museumskunde*, **10**) ed. Grote, A. (Leske & Budrich, Opladen) pp. 11-17.

Gudiol, J. [1971] *Goya y Lucientes, Francisco José de, biography, analytical study and catalogue of his paintings*. Vol. **1-4** (Poligrafa, Barcelona).

Hauser, S. E. [2007] "Adolph von Menzel, «Das Eisenwalzwerk (Moderne Cyklopen)»," in *Ferrum* **79**, 121-132.

Hemken, K.-U. [2008] "Befreiungsästhetik. Die Gruppe »Kölner Progressive« und die Klassische Avantgarde," in *Form und Gesellschaft. Symposium zur Ausstellung »köln progressiv 1920-33«*, eds. Friedrich, J., Gülicher, N., & Roth, L. (Museum Ludwig, Köln, & VG Bild-Kunst, Bonn) pp. 67-77.

Hirdt, W. [2002] *Bildwelt und Weltbild. Die drei Philosophen Giorgiones* (A. Francke, Tübingen).

Hoppe, B. [1994] "Kunstkammern der Spätrenaissance zwischen Kuriosität und Wissenschaft," in *Macrocosmos in Microcosmo. Die Welt in der Stube. Zur Geschichte des Sammelns, 1450 bis 1800* (*Berliner Schriften zur Museumskunde*, **10**) ed. Grote, A. (Leske & Budrich, Opladen) pp. 243-263.

Hoppe, B. [1996] "Die Vernetzung der mathematisch ausgerichteten Anwendungsgebiete mit den Fächern des Quadriviums in der Frühen Neuzeit," in *Der "mathematicus". Zur Entwicklung und Bedeutung einer neuen Berufsgruppe in der Zeit Gerhard Mercators* (*Duisburger Mercator-Studien*, **4**) ed. Hantsche, I. (Dr. N. Brockmeyer, Bochum) pp. 1-33.

Hoppe, B. [1997] "Structural Development of Experimental Methods in the Researches on Gas-Exchange and Photosynthesis," in *Biology Integrating Scientific Fundamentals* (*Algorismus*, **21**) ed. Hoppe, B. (Institut für Geschichte der Naturwissenschaften, München).

Hoppe, B. [1998] "Sublunar/translunar," in *Historisches Wörterbuch der Philosophie*, Vol. **10**, eds. Ritter, J. & Gründer, K. (Schwabe & Co., Basel), col. 477-481.

Hoppe, B. [2000] "Naturforschung im Stil Humboldts in der Entwicklung der Naturwissenschaften," in *Die Dioskuren*, Vol. **II** (*Abhandlungen der Humboldt-Gesellschaft*, **16**) eds. Haberland, D. et al (TZ-Verlag, Mannheim).

Hütt, W. [1981] *Adolph von Menzel* (Vollmer, München).

Johanek, P. [1986] "Klosterstudien im 12. Jahrhundert," in *Schulen und Studium im sozialen Wandel des Hohen und Späten Mittelalters* (*Vorträge und Forschungen*, **30**) ed. Fried, J. (Jan Thorbecke, Sigmaringen) pp. 35-68.

Johnson, U. E. [1964] (ed.) *Das 20. Jahrhundert. Eine Auswahl der besten Meisterzeichnungen von 1900 bis 1940* (English edition by Shorewood Publishers, New York; German translation by Kahr, B.; Licence edition by the Bertelsmann Lesering, Gütersloh).

Kandinsky, W. [1912] *Über das Geistige in der Kunst* (Piper, München).

Kandinsky, W. [1926] *Punkt und Linie zur Fläche. Beitrag zur Analyse der malerischen Elemente* (*Bauhausbücher*, **9**) (Langen, München).

Keim, F. [2009] *Die Entdeckung der Jupitermonde 105 Jahre vor Galileo Galilei. Eine Studie zum Heliozentrismus in der Frühen Neuzeit* (Peter Lang, Frankfurt am Main).

Klemm, F. [1954] *Technik. Eine Geschichte ihrer Probleme* (*Orbis Academicus*, **II/5**) (Karl Alber, Freiburg im Breisgau).

Klemm, F. [1973] "Geschichte der naturwissenschaftlichen und technischen Museen," in *Deutsches Museum, Abhandlungen und Berichte*, **41 no. 2**, eds. Auer, H. & Berninger, E. H. (R. Oldenbourg, München) pp. 3-51.

Klonk, C. [2003] "Science, Art, and the Representation of the Natural World," in *Eighteenth-Century Science* (*The Cambridge History of Science*, **4**) ed. Porter, R. (Cambridge University Press, Cambridge) pp. 584-617.

Köhn, R. [1986] "Schulbildung und Trivium im lateinischen Hochmittelalter und ihr möglicher praktischer Nutzen," in *Schulen und Studium im sozialen Wandel des Hohen und Späten Mittelalters* (*Vorträge und Forschungen*, **30**) ed. Fried, J. (Jan Thorbecke, Sigmaringen) pp. 203-284.

Krafft, F. [1980] "Artes mechanicae," in *Lexikon des Mittelalters*. Vol. **1**, ed. Avella-Widhalm, G. et al (Artemis, Zürich) col. 1063-1065.

Krapinger, G. [1999] "Martianus Capella," in *Der Neue Pauly. Enzyklopädie der Antike, Altertum*, Vol. **7**, eds. Cancik, H. & Schneider, H. (J. B. Metzler, Stuttgart) col. 961-963.

Kuchenbuch, L. [2001] "Sklaverei," in *Der Neue Pauly. Enzyklopädie der Antike, Altertum*, Vol. **11**, eds. Cancik, H. & Schneider, H. (J. B. Metzler, Stuttgart) col. 621-633.

Liddell, H. G., Scott, R. & Jones, H. S. [1958] *A Greek-English Lexicon*, 2nd edition (Clarendon Press, Oxford).

Lindgren, U. [2004] *Die Artes Liberales in Antike und Mittelalter. Bildungs- und Wissenschaftsgeschichtliche Entwicklungslinien* (*Algorismus*, **8**, Neuausgabe) (Dr. Erwin Rauner, Augsburg).

MacGregor, A. [2007] *Curiosity and Enlightenment* (Yale University Press, New Haven).

Mauriès, P. [2002] *Das Kuriositätenkabinett* (German translation by Vogel, S. & Ferstl, R. of »Cabinets of Curiosities«, Thames & Hudson Ltd., London, 2002) (DuMont Literatur und Kunst Verlag, Köln).

Meyer, G. [1886-1887] *Die sieben freien Künste im Mittelalter* (*Jahresbericht über die Lehr- und Erziehungsanstalt [...] für das Jahr* **1885/86 und 1886/87**) (Gebrüder Benzinger, Einsiedeln).

Quiccheberg, S. [1565] *Inscriptiones vel Tituli Theatri Amplissimi [...]* (Adam Berg, München).

Renger, K. [1982] *Graphik in Holland* (Staatliche Graphische Sammlung München, München).

Roth, H. [2000] (ed.) *Der Anfang der Museumslehre in Deutschland, das Traktat "Inscriptiones vel Tituli Theatri Amplissimi" von Samuel Quiccheberg; lateinisch–deutsch* (Akademie-Verlag, Berlin).

Roth, L. [2008] "Malerei als Waffe," in *köln progressiv 1920–33, seiwert – hoerle – arntz, Katalog zur Ausstellung im Museum Ludwig,* ed. Roth, L. (Verlag der Buchhandlung Walther König, Köln) pp. 15-132.

Salley, V. [2003] *Albrecht Dürer. Kunststück Natur. Tier- und Pflanzenstudien* (Prestel, München, Berlin).

Sallmann, K. [2000] "Plinius [1]," in *Der Neue Pauly. Enzyklopädie der Antike, Altertum*, Vol. **9**, eds. Cancik, H. & Schneider, H. (J. B. Metzler, Stuttgart) col. 1135-1141.

Sallmann, K. [2002] "Varro [2]," in *Der Neue Pauly. Enzyklopädie der Antike, Altertum*, Vol. **12/1**, eds. Cancik, H. & Schneider, H. (J. B. Metzler, Stuttgart) col. 1130-1144.

Stahl, W. H. [1970] "Aristarchus of Samos" in *Dictionary of Scientific Biography*, Vol. **1**, ed. Gillispie, C. C. (Charles Scribner's Sons, New York) pp. 246-250.

Ullrich, W. [2005] *Was war Kunst? Biographien eines Begriffs* (Fischer Taschenbuch Verlag, Frankfurt am Main).

Vries, D. J. de [2007] "Begeistert von Rembrandt, von seinen Häusern und von der Erinnerung," in *Das Dürer-Haus. Neue Ergebnisse der Forschung* (*Dürer-Forschungen*, **1**) eds. Großmann, G. U. & Sonnenberger, F. (Verlag des Germanischen Nationalmuseums, Nürnberg) pp. 173-192.

Whitehead, P. J. P. & Boeseman, M. [1989] *A Portrait of Dutch 17th Century Brazil. Animals, Plants and People by the Artists of Johan Maurits of Nassau* (*Koninklijke Nederlandse Akademie van Wetenschappen, Verhandelingen Afd. Natuurkunde, Tweede Reeks, deel* **87**) (North-Holland, Amsterdam).

Wirth, I. [1965] *Mit Adolph Menzel in Berlin* (*Bibliothek des Germanischen Nationalmuseums Nürnberg. Zur deutschen Kunst- und Kulturgeschichte*, **25-26**) (Prestel, München).

3

Science and Art: A Philosophical Perspective

Guo-Sheng Wu

The original source of modern science is Greek liberal humanities. Modern art is something called fine art which is the heritance and development of ancient ideal of free arts. Freedom is the common essence of both science and art. As soon as their Greek mind emerged, the two immediately recognized each other as brothers who were separated from and lose touch with each other for a long time. This is why science and art know each other very well while they are two different departments of human culture.

3.1 Introduction

The distinction between science and art is clear. However, there are also many famous scientists, artists and humanists who emphasize the similarities between science and art and their internal intrinsic association. Science and art have great differences, since the creation of aesthetic object is the aim of Art but the object of science is not created by Science. Yet, many scientists emphasize much on the role of aesthetics in scientific research. Thus it raises a question: What kind of interrelationship it is that is between science and art?

3.2 Origin of "Science"

Here, the origin of the word "science" is discussed. To modern people, the main difference between science and art appears to be that the two have different groups of practitioners, one called the *Science Community*; the other, the *Art Circle* of artists. However, we must know that the clear division of scientists and artists as different professional groups has a history of only about three or four hundred years.

Science became a profession in the first half of the 19th century, signified by the creation of the word "scientist" by William Whewell (1794-1866) in 1833. Before that time, scientists were usually called "natural philosophers" and they had various occupations such as university professor, government officer, factory owner, drug seller, and gardener. The appearance of the word "scientist" showed that *science* had already separated from the matrix of philosophy to be an independent social entity.

When "scientists" replaced "natural philosophers", "science" also replaced "natural philosophy" to be a well-known word at the same time—a sign of the independence of science and the professionalization of scientists. The English word "science" *usually* refers to "natural science", which reflects this historical fact of self-restricting and self-delimiting.

Although science was separated from philosophy, it also bore the heritage of ancient scientific tradition from ancient Greece; basically speaking, philosophy also belongs to this ancient tradition. In etymology, the word "science" was derived from the Latin word "*scientia*", while *scientia* was the translation of the Greek word "*episteme*". Throughout the entire Middle Age, *scientia* had the same usage as *episteme* in Greek texts. *Episteme* represents a tradition of pursuing for certain and universal knowledge. "The eternal order of things" and "the universal truth" are the exact meaning of *episteme*.

It is very interesting that "science", a fixed term in early 19th century, does represent some specialized branches of "universal knowledge", while the word itself contains the ideal of "universal knowledge" in its etymology. The Greek ideal of "universal knowledge" is being still insisted upon by some professionalized scientists, especially theoretical physicists. This "universal knowledge" is a great creation by the ancient Greeks.

What is "*episteme*"? In *Cratylus,* Plato mentioned this word like this: it "seeming rather to signify stopping the soul at things than going round with them". [1] That is to say, it is in a stable state. To the Greeks, knowledge must be certain and eternal; otherwise it would be disputable and changeable ("*doxa*").

The pursuit of the certainty of knowledge had a root in the soil of Greek democratic politics which was full of polemics. Can we arrive at an end in these discussions, debates, and arguments? Is there much to be said on both sides? Socrates turned against the sophists who took debates just for debates' sake, leading the arguments to a much higher goal in which knowledge is certain and universal, thus with a power of restrain. This power of restrain does not come from your skilled rhetoric, but from knowledge it*self*.

Knowledge has a "self"; that is the essential reason why knowledge is certain and universal. The "selfness" of knowledge is represented in Plato's "Ideal World". The Ideal World comes to be by itself, runs by itself and unfolds by itself. All real knowledge must be the knowledge of the ideal world. The Ideal World is the so-called "world of essence" or "internal world". It is "self-sufficient", "self-dependent", and "self-disciplined"; it thus has a power of restrain.

Therefore, the Greek ideal of universal knowledge is embodied in the pursuit of the ideal world which comes to be by itself and unfolds by itself. To the Greeks, freedom means to comprehend this transcendency and internality, i.e., to grasp the pure knowledge or the absolute knowledge. The principle of internality is just the principle of freedom.

Aristotle said, "Therefore, since they took philosophy in order to escape from ignorance, evidently they were pursuing science in order to know, and not for any utilitarian end… as the man is free, we say, who exists for his own sake and not for another's, so we pursue this as the only free science, for it alone exists for its own sake."[2] Aristotle divided all knowledge (or science) into three categories: theoretical science, practical science, and productive science, among which theoretical science is the highest one as the real-and-free knowledge. *The ideal of universal knowledge since Greek times, in fact, is just the ideal of free science.*

---

[1] Plato, *Cratylus* 437A [Hamilton & Cairns, 1989].
[2] Aristotle, *Metaphysics*, 982b20-28 [McKeon, 2001].

In his *Nine Books of Discipline*, Roman writer Varro (116-27 BC) put forward nine subjects to cultivate the dignity of Romans; the nine subjects are grammar, rhetoric, logic, arithmetic, geometry, astronomy, music, medicine, and architecture. He gave them a name in total as "liberal arts" (*Artes liberales*) because they were designed for free men, not slaves. In the 5th century, Martianus Capella removed the last two, and thus formed the well-known "seven arts" in the Middle Age, among which the first three subjects were called "trivium"; the last four, "quadrivium". In fact, modern science just inherited the ideal of free science from Greek times by developing the quadrivium.

The invention of the word "science" as symbol of professionalization and specialization was once rejected by the top scientists at that time because they felt that they were put outside the great tradition of universal knowledge and free science by adopting this word. Even until the 1890s, Thomas Huxley, Lord Kelvin, and Lord Rayleigh still refused to call themselves "scientists". This word became popular only in the 20th century. However, some scientists still want to persist in their ideal of Greek science by other ways. One way is to return to the ideal of free science by connecting science with art, emphasizing the aesthetics factor in scientific discoveries.

3.3 Origin of "Art"

It is very interesting that when Whewell invented the word "scientists", he was imitating the word "artists". It indicates that there was a professionalized group called "artists" before the existence of "scientists". However, modern art[3] just sprung up in the Renaissance. More specifically speaking, the modern concept of "art" was not established until the 18th century.

What exactly is it that is called "art" in a modern sense? Modern "art" is the so-called "fine arts", which is in contrast to *technics*. Ernst Gombrich (1909-2001) [1993], a historian of art, said it very well: "If

---

[3] The term "modern art" in this chapter refer to art that appeared since the Renaissance (14th-17th centuries), and should not be confused with Modern Art that appeared after Cézanne during the last century.

something itself became the aim, then we have the right to call it art." Another historian of art, Erwin Panofsky (1892-1968), contented that an artwork could be called an artwork just because it has the pure meaning of aesthetics, when the concept of so-called aesthetics just appeared. He [1955, p. 11] said, "When do this, to express it as simply as possible, when we just look at it (or listen to it) without relating it, intellectually or emotionally, to anything outside of itself. ... Only he who simply and wholly abandons himself to the object of his perception will experience it aesthetically." He [1955, p. 12] gave an example to show the difference between artworks and non-artworks:

> If I write to a friend to ask him to dinner, my letter is primarily a communication. But the more I shift the emphasis to the form of my script, the more nearly does it become a work of calligraphy; and the more I emphasize the form of my language (I could even go so far as to invite him by a sonnet), the more nearly does it become a work of literature or poetry."

It is very clear that modern art emphasizes a kind of non-pragmatic, pure aesthetic meanings. Of course, art is an artificial creation. However, it has an essential difference from the normal artifact that it is created just for the aesthetic purpose.

When did this conception of art come about? Historians of art generally agree that "art as a noble and divine concept appeared in the 18th century." In 1746, an French priest Abbe Charles Batteux (1713-1780) published the book *Les Beaux-Arts Réduits à un Même Principe* (*The Fine Arts Reduced to a Single Principle*) in which he divided art into three categories:

1. Fine arts with a purpose for pleasure, including five sorts: sculpture, painting, music, poetry, and dance;
2. Artificial arts with a purpose for utility;
3. Arts with mixed purposes for pleasure and utility, including rhetoric and architecture.

Batteux's division definitely separated fine arts from the mechanical arts (see below); consequently, it was taken as a sign for the establishment of the modern- art system. However, in reality, the establishment of the modern-art system was a gradual evolving process; it was not made once and for all.

The word "Art" derives from "*ars*" in Latin and "*techne*" in Greek. *Techne* has a methodological meaning as "proficiency" and "acquaintance", which could be used to refer to any human manufacture and human action, including what we called today arts and handicrafts: not only including sculpture, painting, music, poetry, and dance, but also vintage, shoe making, farming, and cooking. It can even be applied to *episteme*, but with a different meaning: *Techne* means manual making and operation while *episteme* means "disinterested understanding". In this sense, *Techne* is at a level lower than that of *episteme*. However, at that time, there was no distinction between art and techniques, as we have today. The five modern categories of art were not regarded as a unified term. In early Greek, the nine creative activities under the nine goddess of muse are closer to today's taxonomy of art: epic poetry, history, lyric poetry, music, tragedy, choral poetry, dance, comedy, and astronomy. They are clearly different from the so-called art in a modern sense: history and astronomy are surely not included in today's art, while today's visual arts such as sculpture and painting are not included in the nine categories.

The Latin word *ars* generally inherited the usage of the Greek word *techne*. However, the seven liberal arts, which were put forward by Latin writers, showed that *ars* were in a broad meaning of *techne*. In the 12th century, Hugo of St. Victor (1096-1141) added seven mechanical arts—lanificium (wool spinning), armature, navigation, agricultura, venation, medicina, and theatrica—to the seven liberal arts. The fine arts, as we call them today, are distributed among the seven mechanical arts and the seven liberal arts. For examples, architecture, sculpture, and paintings are included in armature.

In the Renaissance, the concepts of science, technique, and art were still in a mingled state without any significant change. Gombrich [1993, p. 74] said this about Vinci, "This is important, from the historian's point of view; I have had to deal with it sometimes. When I was writing about

Leonardo, for instance: There is all this boring stuff about how he combined art and science, because he did not have these notions: 'art' meant skill and 'science', knowledge." However, it is necessary to notice that the visual arts (such as painting, sculpture, and architecture) emerged surprisingly in the Renaissance and made great achievements. The artists, represented by Leonardo da Vinci (1452-1519), consciously endowed the new visual arts with great ideals, trying hard to separate them from the traditional mechanical arts and putting them into the rank of the traditional liberal arts.

In order to enter the rank of liberal arts, the new generation of visual artists consciously moved closer to a kind of traditional liberal arts, i.e., mathematics. Mathematical education became a primary training course for artists at that time. New visual artists used mathematical methods, represented by the principles of perspective, to reorganize their art space as the new scientists reorganize their physical space (Fig. 3.1). It was during this period that the artists, as we call them today, played a significant role in the process of creating new science. Da Vinci was just an outstanding example. For them, science and art had not separated from each other yet.

Fig. 3.1. A Renaissance artist in the process of combining geometry and art. [Woodcut from Albrecht Dürer's *Underweysung der Messung mit dem Zirckel und Richtscheyt* (Nuremberg, 1525).]

The Italian Giorgio Vasari (1511-1574) put forward the concept of "Arti del disegno" (design art), grouping the visual arts—such as painting, sculpture, and architecture, into a unified art category. He also led a group of painters, sculptors, and architects to break away from their former guild organizations and established the first modern art college.

In 1690, the French Charles Perrault (1628-1703) proposed to use "fine art" to replace the concept of traditional "liberal arts" in his article "Art Gallery". He listed eight subjects of Art in total: rhetoric, poetry, music, architecture, painting, sculpture, optics, and mechanics. In these eight subjects of art, although optics and mechanics were retained, traditional rational disciplines such as arithmetic, geometry, and astronomy were removed. It shows a conscious effort to make distinction between science and art.

The formal divorce of science and art was announced by Batteux through his book in 1746 (see Section 3.3). In fact, the Encyclopedists of the 18th century put science and art side by side, classifying *all* human knowledge into these two categories. Since the early Greek time, the fine art and the free science experienced a series of divorce and marriage before finally shaped into two clearly distinct careers, two different exploring fields for humans' intellectual elites.

3.4 "Freedom" as the Common Nature of Science and Art

The historical review above shows that modern art had experienced a period during which it took an initiative to approach liberal arts and shape its pure aesthetic character. The real art is the same as the real-and-free science that they are both beyond any pragmatic purpose. Freedom is the common nature of science and art. When the Greek spirit resounds in their bodies, they recognize each other at once as brothers missing for years.

Ernst Cassirer (1874-1945) [1944] said it very well: "To the Greek mind beauty always had an entirely objective meaning. Beauty is truth; it is a fundamental character of reality. If the beauty which we feel in the harmony of sounds is reducible to a simple numerical ratio it is number that reveals to us the fundamental structure of the cosmic order." The cosmo which represents the order is perfect and understandable. The seeking for knowledge and aesthetic meaning was unified by the early Greeks.

The Renaissance artists represented this unification in the body art. The research of the theory of human-body ratio became a precondition to make any art creation, which was the embodiment of the harmonious ideal of Neoplatonism in the microcosm of the human body. Astronomers research the harmony of macrocosm, while the artists research the harmony of microcosm. Panofsky [1955, p. 91] once said, "The proportions of the human body were praised as a visual realization of musical harmony; they were reduced to general arithmetical or geometrical principles (particularly the "golden section," to which this period of Plato worship attached a quite extravagant importance)..." It was for this reason that the Renaissance artists actually engaged in the construction of modern mathematics and modern mechanics.

The classic art is sacred and divine, not just an unimportant entertainment for life ornament. The classic science is also sacred and divine, not just a method which can be applied to solve sorts of actual problems. Today's artists would be the same as scientists when science and art return to their original state, the state of free science and free art. When that happened they would surely enter a realm in which "Beauty is Truth".

References

Cassirer, E. [1944] *An Essay on Man: An Introduction to a Philosophy of Human Culture* (Yale University Press, New Haven) p. 266.

Gombrich, E. [1993] *A Lifelong Interest: Conversations on Art and Science* (Thames & Hudson, London).

Hamilton, E. & Cairns, H. (eds.) [1989] *Plato: The Collected Dialogues* (Princeton University Press, Princeton) p. 471.

McKeon, R. (ed.) [2001] *The Basic Works of Aristotle* (Random House, New York) p. 692.

Panofsky, E. [1955] *Meaning in the Visual Arts: Papers in and on Art History* (Doubleday, New York).

4

Neuroarthistory: Reuniting Ancient Traditions in a New Scientific Approach to the Understanding of Art

John Onians

There is a long tradition of using neuroscience to understand art. From Aristotle to Baxandall scholars who we might think of as humanists have turned to science for help. This, however, became unfashionable at the end of the twentieth century and the practice is only now being revived, reuniting ancient traditions. This chapter presents some of the latest knowledge of the brain that is revolutionizing the field and illustrates its application. It also argues that the issue of whether humanistic and scientific traditions can be reunited is crucial to the future of intellectual enquiry.

4.1 Introduction

Art as part of science has a long tradition, dating back at least to Aristotle, one which unfortunately was broken after the Roman Empire. However, today the importance of the link between the fields is recognized by more and more humanists.

4.1.1 *A Long Tradition*

Two and a half thousand years ago in ancient Greece Lui Lam's vision of the arts as an aspect of science [Lam, 2011] would have been normal. This was especially true in the school of Aristotle, who was happy, unlike Plato, to consider humans as just another type of animal, and who wrote studies of politics and drama, for example the *Politics*, which are in many ways as scientific as his studies of biology, such as the *Parts of*

Animals. But, after the Fall of the Roman Empire, science and art drifted apart, only being brought together occasionally by figures like Leonardo da Vinci. It was, however, in the nineteenth and twentieth centuries that the separation was made official with the recognition in universities of the sciences and the humanities as fields distinct both in their materials and their methods.

The negative consequences of this division were first fully brought out in C. P. Snow's analysis of the situation in terms of the currency of "two cultures". Subsequently the division only became more entrenched, with scientists, who stressed the universality of scientific laws, opposed to humanists, who argued that the triumphs of human creativity in fields such as literature, music and art were the products of the free, individual human mind. This polarization further intensified when so-called post-modern theory argued that nothing cultural could be studied scientifically because it was all "socially constructed". Fortunately, though, there have recently been signs that things are changing. Most significantly, more and more leading thinkers in the humanities, including specialists in the study of art, are coming to see major aspects of such human behaviors as subject to the laws of nature, and understandable in biological rather than cultural terms. And, perhaps surprisingly, this is equally true both of old fashioned positivists and trendy postmodernists.

4.1.2 *Humanists Who Convert to Science*

Among the old fashioned positivists there are no more striking examples than Sir Ernst Gombrich, who earned his knighthood above all for his work on the art of the Renaissance and its successors, and Sir John Boardman, who earned his for his work on the art of Classical Greece. Both spent a lifetime celebrating the freedom of individual artists and the freedoms of the societies in which they flourished, but in their last major books, published when they were in their late sixties and seventies, they present a quite different view, giving nature not an incidental but central role in the shaping of culture and especially artistic culture. Gombrich in *The Sense of Order* (1976) described the making of ornament as rooted in man's "biological inheritance" and even told an interviewer in 1992, that "my approach is always biological" [Gombrich, 1993] while

Boardman in *The World of Ancient Art* (2006) declared that the story he told revealed the extent to which "climate and the environment completely dictated the development of culture" [Boardman, 2006]. Such views need to be taken particularly seriously, because they are at odds with everything those authors had written earlier. Evidently late in life the penny had finally dropped that man was not the god-like free agent they had celebrated earlier, but very much a part of nature and subject to nature's laws.

The examples among the postmodernists are just as striking. W. J. T. Mitchell, one of the founders of the disciplinary framework, Visual Studies, within which postmodern theory has particularly thrived, has been increasingly presenting himself as open to a more naturalist perspective. Thus, in his 1994 article on "What is visual culture?" he cautions that we should not "ignore scientific accounts of vision and imaging, whether addressed to the 'natural' construction of the eye and the visual system in its relation to the other senses, or to the science and technology of imaging" [Lavin, 1994]. And more recently he has admitted that visual culture is not just about the social construction of the visual but the visual construction of the social, bolstering this with the statement that "This version of visual culture understands itself as the opening of a dialogue with visual *nature*. ….It does not content itself with victories over 'natural' attitudes and 'naturalistic fallacies', but regards the seeming naturalness of vision and visual imagery as a problem to be explored, rather than a benighted prejudice to be overcome. In short, a dialectical concept of visual culture cannot rest content with a definition of its object as the 'social construction of the visual field' but must insist on exploring the chiastic reversal of this proposition, *the visual construction of the social field*. It is not just that we see the way we do because we are social animals, but also that our social arrangements take the form they do because we are seeing animals" [Holly & Moxey, 2002].

Mitchell has not had the courage to follow through on his argument, but another erstwhile apostle of postmodern theory, Norman Bryson, has. As he says in his introduction to a new book by Warren Neidich, *Blow-Up: Photography, Cinema and the Brain* (2003), the neuroscientific approach to art adopted by its author offers an entirely fresh "paradigm for thinking through cultural history and the philosophy of the human subject" [Neidich, 2003]. And he goes on to assert that his earlier creed of poststructuralism denies "emotion, intuition, sensation—the creatural life of the body and of embodied experience", features that are fundamental to the human experience of meaning and that implicitly align humans with the animal world. As he says, his earlier models, the heroes of postmodernism, Wittgenstein, Freud, Foucault and Derrida, all made the mistake of thinking of meaning only in verbal terms. Bryson thus contemptuously relegates at least three of the main streams of recent thought—Wittgensteinian philosophy, deconstruction and psychoanalytic theory—to the domain what he calls the purely "clerical". This is because they all give too much importance to words. The meaning of an apple is not captured by the word "apple" but by "the simultaneous firing of neurons and axons" in the brain [Neidich, 2003]. He wants to abandon semiotics for neuroscience. What he looks forward to is a study of art rooted in an understanding of something treated as a fiction by many postmodernists, human nature.

Bryson does not indicate how he personally would apply neuroscience to the study of art. Indeed, he reveals the limitations of his knowledge when he talks of the firing of neurons and axons, apparently unaware that an axon is only a part of a neuron. But his positive reference to hard science is an indication of the direction in which he thinks we should go, and that is the way I have been going myself for the last seventeen years. He wants us to use science to help the humanities. I have attempted to go further, trying to make the humanities more scientific.

4.2 From Art History to Neuroarthistory

Here we describe the rationale behind the change from art history to neuroarthistory and my personal journey related to it.

4.2.1 *From Art History to World Art Studies*

One reason I wanted to go further was because my starting point was more like that of a scientist than a humanist. Humanists typically start from a single object, a book or a work of art, an artist or writer, the culture of a town or a country and try to gather information that relates to it. Usually this information comes from books, and these are typically written in a recognized language. That was the way I myself had worked until 1992. It was then that I persuaded my colleagues to change the name of our department from School of Art History to School of World Art Studies. Although I did not realize it fully at the time, this had massive implications for the way we approach art [Onians, 1996]. Art history operates with small data sets, such as Renaissance art, that are place- and period-bound. To study art as a worldwide phenomenon is to work with a data set that is infinitely larger. For a start it requires considering all the art ever made anywhere on the globe during at least 40,000 years. It also requires a redefinition of art. Art historians typically think of the category art as being made up of paintings, sculptures and buildings, because those are typical of Europe. But once you go outside Europe the most important object might be a piece of Arab calligraphy, a grass skirt from Tahiti or a mask from Africa. World Art Studies must address them all. The disciplines I had used until then were no help. Indeed, none of the humanities were. The art of Europe could be studied using the approach of the typical humanist. World art could not. World art is more like "flora" or "fauna", both of which categories inherently refer to the whole world. World art needs to be considered as a vast single field, as nature is considered by naturalists. Darwin could never have understood finches if he had stayed at home. He needed to travel and see not just finches, but every other bird, not to mention all the food they might eat and the full range of their ecologies, the climate, geology etc. I felt I had to do the same, if I was to understand art. I had to consider all of art and I had to consider the contexts of its production and consumption. I also needed to develop a completely new approach. It was no longer possible to relate the art to information in a few languages. I had to relate it to something more fundamental. What should that be?

4.2.2 *From the Conscious to the Unconscious*

It took me some time to realize that there had been great losses associated with the concentration on verbal information. Giving prominence to words means giving prominence to the conscious mind, but much of the mind's activity is not accessible to consciousness. If we only want to study culture the way we have done in the past, we can stay with words, and use them to give us access to conscious mental activity. But if we want to treat humans as animals and see culture as an aspect of nature, trying to understand mental activity that is unconscious, we have to find another way of accessing the mind. In the case of art, the inadequacy of language as a means of access to mental activity is particularly clear. A novelist or philosopher may plan a work using words, an artist generally does not. Many aspects of what he or she makes are the product of mental processes which relate primarily to the operation of the eye and the hand, and have nothing to do with words. Indeed because the processes are ones of which the maker is usually completely unaware, he or she cannot even express them verbally if he or she wants to. My problem was how to get access to such hidden processes.

It was then that I realized that both our unconscious relation to our environment and the planning, making—and for that matter—our response to art, take shape in our brain. If I could find out how the brain worked, I thought, I might get access to a whole range of mental processes, which had previously been more or less completely ignored, although their importance had long ago been recognized by philosophers such as Locke and Hume. *And that is where neuroscience comes in.* In order to get some access to those unconscious mental processes, I decided in 1992 to find out as much as I could about the brain's workings. Although I did not realize it at the time, I was turning myself from a humanist into a scientist. What I once considered as culture, the product of the conscious mind of a god-like man, I now thought of in natural terms. In effect I was becoming an ethologist and a neuroscientist. My total data set was a particular behavior of a particular species, *Homo sapiens*, and the principal tool I was using to understand it was an aspect of science, knowledge of the brain.

4.2.3 *Advantages of a Scientific Approach*

At first I thought that there would be costs to my new approach. Scientists tend to generalize and they do not like talking about the individual. Ethologists look for species-wide behaviors that are driven by genes. Neuroscientists, too, try to understand behavior in terms of universal processes. But even as I followed these fields during the last two decades, both genetics and neuroscience began to embrace the individual. The analysis of the genome has made it possible to look at genetic inheritance in terms not of a species but of a single organism, and an understanding of neurons has revealed the extent to which our neural make-up is unique to ourselves, being shaped by every single experience we have ever had. The new scientific knowledge, then, provides me with a tool which not only gives me access to previously invisible human mental activity species-wide, but also gives me access to that prime concern of the humanist, the individual mind. Far from losing sight of the individual by turning to science, I was discovering that the single most important benefit of a neuroscientific approach is that it offers a much more complex model of the mind than anything available in the humanities.

This is because it is a model which acknowledges that we are all born with a brain made up of a hundred billion neurons, that each of these can have up to a 100,000 connections with other neurons, that these connections are constantly being made and falling away in response to changes in our sensory experiences, emotions, thoughts and actions, and that those connections function thanks to a constantly changing supply of different neurochemicals, among the most important of which is the pleasure-giving dopamine, which cause us to repeat actions that we enjoy. Not only is it a model that acknowledges all aspects of our inner life, from the most visceral to the most rational, and from the most emotional to the most Olympian, but for the first time it allows us to track in extraordinary detail how that inner life changes, not just from year to year, but from day to day, from minute to minute. A neural approach to any aspect of the humanities requires us to take into consideration many more sides of human life than we usually do. Treating humans as animals in fact gives us a much richer understanding of the resources which make them god-like. By adopting the approach of a scientist I was able to get ahead of the humanist in precisely those areas in which they are proudest of their expertise.

That at least is my big claim, and a little later I will try and substantiate it, but, before I do so, I should say something about the extent to which this approach falls short of the strictest definition of science. Although I am using scientific knowledge I am not generating it. I am not running experiments, doing statistical analysis or publishing refereed papers in *Nature*. My work is not scientific, only based on science. Besides, the mental activity to which I have gained access is only of one type, that of which we are not conscious. Paradoxically, I have no new direct access to conscious mental activity. If I want to understand conscious decision-making or the consequences of reading a book or talking to someone, I am still best advised to use the same tools I have always done, words. My only advantage is that I may now be able to judge when conscious mental activity is driven by some other activity that is unconscious. So, just as I need to keep in touch with ethologists and neuroscientists, I also need to keep friends with people who know the languages of the people whose behaviors I am investigating. Indeed, I need to know enough about languages to be able to evaluate what those people are telling me, just as I need to know enough about neuroscience and genetics to evaluate scholarship in those fields.

4.3 Neuroscientific Tools: Neural Plasticity and Neural Mirroring

So, having read these broad claims for my project of neuroarthistory, you are probably anxious to find out how it works. What is the neuroscientific knowledge which I apply to my data set of world art? The first is that relating to *neural plasticity*, a phenomenon whose importance emerges every day more clearly, changing our whole view of the brain [Huttenlocher, 2002]. It thus used to be thought that the brain was a sort of machine that, once formed in childhood by a process of biological programming, remained more or less stable. It is now realized that it is an organ in a state of constant change throughout our life, with connections between its hundred billion neurons being formed and falling away in response to changing experiences, endowing each individual with networks supporting specific inclinations and skills. The most important aspect of neural plasticity for those studying art is the way our neural networks are reshaped by our visual experience. Each time we look with attention at anything, such as a plant, an animal or a man-made object, or a feature of the general environment, such as the sky or the soil, new connections are formed between the neurons involved, connections that improve our ability to see that object and give us a preference for looking at it or any other object that shares similar attributes. This means that if we know, or can reconstruct, what an individual at a particular place and time was looking at with particular attention, whether it was part of their natural environment or was man-made, we can become aware of an important factor potentially influencing their aesthetic choices, and this is true whether that person was a maker of art, a patron of art or a viewer of art.

A second area of knowledge that is particularly useful when studying art is that relating to so-called "mirror neurons". It was known for a long time that we have separate groups of neurons in our brain that fire when we move a particular limb, such as a hand or a finger. Then, in the early 1990s, an Italian team of neuroscientists discovered that, in our close relatives, macaque monkeys, some of these neurons also fire when one individual merely observes another moving that limb, although they themselves make no movement [Rizzolatti, 1996]. Soon it was realized that such *neural mirroring* is even more important for human beings. For example, these same mirror neurons help us quite unconsciously to learn highly specific movements. And since these neurons are also associated with an ability to understand why the action is performed, they are also an aid to our understanding of what goes on in other people's minds. These neurons lie behind human beings' astonishing ability to imitate each other without any formal instruction being involved, as when a child learns movements from its parents. For art historians they help us to understand many aspects of art-related behaviors. For example, they make us aware of how the bodily dispositions and movements of an artist when at work may be influenced by the movements of those individuals to whom that artist or designer has given particular attention.

4.4 Neuroarthistory and Its Applications: Selected Examples

Neuroscience as applied to arts from different periods and locations are presented here, illustrating the applications of neuroarthistory.

4.4.1 *Neuroscience Applied to History: Prehistoric Art*

There are many other areas of neuroscientific knowledge that are of value to the art historian, but it is these two that I want to particularly exploit here. And there can be no better point to begin doing this than with the very origin of representational art, the paintings in the Cave of Chauvet in Southern France (Fig. 4.1). The challenge of explaining these works is especially great considering that not only, at 32,000 years old, are they absolutely the earliest known, they are also extraordinarily life-like, being much more naturalistic than most later art, beating not just Lascaux but Leonardo. There is nothing to match the way they capture animals running wild, that is not until David Attenborough filmed the Serengeti. They even exploit ¾ perspective (Fig. 4.2) something not seen till Greek art 30,000 years later. No available approaches can explain why this art suddenly appears and why it has these extraordinary attributes. An art history that uses neuroscience, that is neuroarthistory, can.

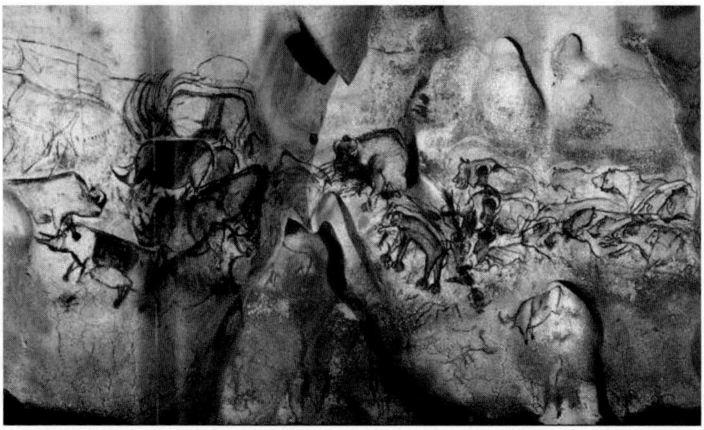

Fig. 4.1. Chauvet Cave, Vallon Pont d'Arc: General view of paintings in niche area.

Fig. 4.2. Chauvet Cave: Cave bear painting.

At several points in the cave marks made by cave bear paws are followed by those made by human hands. Where bear claws have scratched the surface, humans have made engravings (Fig. 4.3, left) and where muddy bear paws have made brown marks, humans have made paintings (Fig. 4.3, right) sometimes employing the same technique, pressing palms covered with pigment on the cave walls. The sight of the two types of marks made by the humans' more powerful neighbors, the cave bears, evidently activated their mirror neurons, causing them unconsciously to imitate the bears in the same way that children unconsciously imitate their parents. And, once they started to mark the walls, the shapes they made were those to which neural plasticity had made them most sensitive. Newly arrived in Western Europe from Africa, close to the edge of an icecap which extended much further south than today, they found themselves at the limit of human habitability. In this extremely cold and hostile environment the things at which they looked with most attention were the animals naturally best equipped to survive in that climate. These were the animals that they looked at with a mixture of admiration and envy. They included bears and lions, rhinoceros and mammoth because those were the creatures whose tool-kits of horns, teeth, claws, fur coats—and, above all, intelligence—they most envied. The principles of neural plasticity, which ensure that the more we look at something with attention the better we will get at seeing it, made it likely that it was the shapes of those same animals that people would tend to see in, or rather project onto, the accidental lines and colors of the cave walls. By the same token, it was the pleasure-giving chemicals that were released when they saw the suggestions of those shapes that led them to complete them using techniques that their neural resources for mirroring had caused them unconsciously to pick up from the cave bears. Working on these images gave them pleasure, and they are so photographic because they went on working on them until they were so close to the image to which their neural resources were tuned, that there was no pleasure to be had by further marking. Neuroscience explains both why they started to make the images and why they stopped.

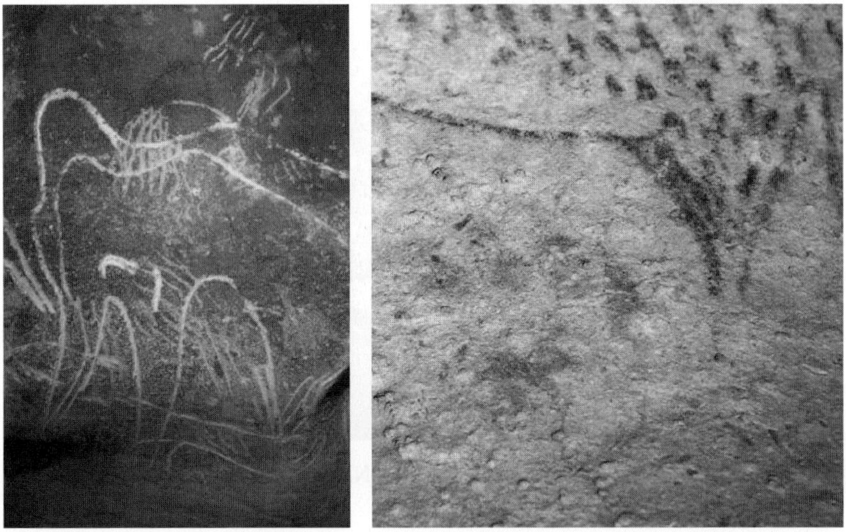

Fig. 4.3. Chauvet Cave. *Left*: Cave bear claw scratching-surface marks and human engravings. *Right*: Cave bear paw brown marks and human hand prints.

But, there are many other limestone caves in France; so why do such works only appear at Chauvet? The answer probably has to do with its location overlooking a unique landscape feature, a rock arch over a dangerous river (Fig. 4.4). Each spring and autumn this arch would have been crossed by a procession of migrating animals and this would have meant that the neural networks of those who watched it would have been exceptionally reinforced. And at the same time the networks involved in viewing the landscape setting with the arch would have been similarly strengthened, which is why when they found a niche which looked like the arch in the wall of the cave they imagined the animals moving over it, and got a particular neurochemical pleasure from completing the shapes not of individual animals, but of a migrating herd (Fig. 4.1). Seeing the niche caused them to imagine the arch and the animals streaming over it, and so this was the scene that they completed.

Fig. 4.4. Vallon Pont d'Arc: Natural rock arch.

4.4.2 *Neuroscience Applied to Geography: Japan and Europe*

Neuroarthistory enables us to understand why art first appears at a particular place and time and it also sheds light on the principles governing the stylistic variations that are such an important aspect of its later history. The earliest representations of the female body that have survived in Europe and in Japan are very different. A typical European figure, the so-called Venus of Laussel, from around 20,000 years ago is large and fleshy and carries a large hard object (Fig. 4.5, left). The Japanese figure, engraved on a small pebble from Kamakuriwa from about 11,000 years ago, is small and flat, with prominent hair and skirt (Fig. 4.5, right). Applying the principles governing neural plasticity and their likely impact on the dopaminergic system, we would expect the difference to relate to differences in the objects to which the makers gave most visual attention, and one of these differences is clear. The inhabitants of Europe at this date, desired and pursued large herbivores which they killed and butchered using large stone tools. In Japan by contrast, the diet was of small mammals, birds and fish, which were often caught by nets, nooses and lines, all made of fibers, which thus became the principal object of the local population's desire. That is why the European gives the woman the attributes of a large herbivore and the Japanese takes delight in her fibrous hair and skirt. The Japanese had looked so intently at fibers that they had developed neural networks that gave them a special interest in them and a neural chemistry that meant looking at them gave them particular pleasure.

Fig. 4.5. *Left:* Venus of Laussel, sculpture. *Right*: Kamakuriwa pebble, drawing.

The currency of these preferences in Japan can also be confirmed by later art, since the same factors lie behind the development of the distinctive ceramic tradition known by the word for string, Jomon. String patterns are found in other early pottery traditions, but only in Japan are they so prevalent as to give a name not only to a pottery style but to a whole period of history. Fibers meant more to prehistoric Japanese because of their ecology and they continued to do so, especially after the arrival of silk technology from the Asian mainland. Indeed the original interest in visible fibers led to the Japanese developing a taste for fibrous textiles which survives to this day in the couture of the country's most famous fashion designers. The continuing importance of fibers in Japan is also illustrated by the prominence of so-called "fiber arts" on the Japanese art scene, a position unknown elsewhere. I also was reminded of the roots of Japanese interest in fibers at the last *Documenta* at Kassel, where the Japanese artist, Ryoko Aoki, had made an image of a woman wearing a skirt in the shape of Mount Fuji (Fig. 4.6). The preferences apparent in the Kamakuriwa pebble are evidently still alive today. And equally revealing of Japanese neural tastes is the reference to Mount Fuji. Only in Japan, where fibers were a long-standing obsession and where the great snow-covered cone dominates the modern capital, would an artist possess neural networks that allowed them to see a flimsy skirt as recalling an extinct volcano.

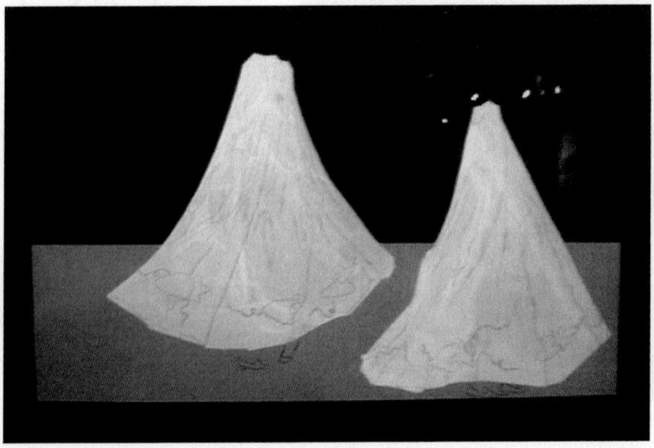

Fig. 4.6. Ryoko Aoki, *Fuji Skirt*, painting.

4.4.3 *Neuroscience Applied to Geography: Athens and Rome*

Reference to volcanoes allows me to suggest how neuroarthistory can also help us to understand regional traditions of architecture. We all know that the supreme expression of the architecture of ancient Greece was the pedimented temple, as exemplified by the Parthenon, while that of Rome was the great domed cylinder of the Pantheon, but no one has asked what the origin of the difference in preferences in the two communities was. Now we can both ask the question for the first time and answer it. The most prominent feature of the environment of both Athens and Rome were the surrounding mountains, and the difference between those mountains is precisely that between the temples. The angular geometry of the ridges of Mount Hymettus anticipate the ridges of the temples on the Acropolis, while the cones of the extinct volcanoes, their craters filled with circular lakes, which ringed the Eternal City, anticipate both the circular shape of the Pantheon and the circular opening with which it is crowned. Looking at the masses of the surrounding mountains formed the neural preferences that shaped the architectures of the rival founders of the European Classical tradition.

4.4.4 *Neuroscience Applied: China and Europe*

The question about the origin of the favorite architectural forms of the Greeks and the Romans only confronted me because I wanted to study art, as scientists study nature, as a global phenomenon. In this sense it was like Darwin's question about the differences in the forms of the beaks of the finches of the Galapagos. Like Darwin I needed a general framework of explanation that could be used to answer similar questions in other contexts. Darwin's framework was the idea of natural selection; mine was the concept of neural adaptation. Darwin did not know the mechanism through which the selection occurred, genetic transmission. I am more fortunate in that much is now known about the mechanisms of neural adaptation, especially the mechanism of neural plasticity.

As Darwin worked he found that some forms and traits, like the peacock's tail and its mating behavior, could not be explained by the basic process of natural selection, and so he had to develop an additional framework of explanation, that of sexual selection, which is really a special form of natural selection. I too have been confronted by artistic phenomena which cannot be explained in terms of neural adaptation, at least not in terms of neural adaptation through plasticity. One such phenomenon is the way the artist holds himself or herself when working. Clearly that cannot be explained through the plasticity of the visual cortex. It can, however, be explained as noted above by another neural mechanism, neural mirroring, which results in the viewer adopting some aspect of the physical pose or movements of the viewed, what I call *neural assimilation*. We have already seen one example of such neural assimilation that between the artists at Chauvet and the cave bears. Their neural mirroring resources led them to mirror the bears' painting activities, resulting in a neurally based assimilation of behaviors.

But if the mechanism is a universal one, we should expect to invoke it to explain other types of artistic deportment, and that is what I have done when looking for an explanation of the origins of the different bodily dispositions of painters in the two most famous of the world's painting traditions, that is those of the oil painters of post-Renaissance Europe and of the ink painters of China. Thus, in Europe, since the sixteenth century, it has been taken for granted that the painter stands before a vertical easel with a brush in one hand and a palette in the other (Fig. 4.7, left), while in China it has been assumed for even longer that the painter sits before a flat sheet of paper or silk and holds a thick brush vertically to drag ink on to it from a pool-like inkwell. There is no doubt that the basic reason why the Chinese sheet is horizontal is because it stops the ink running, but neuroarthistory would look for a single explanation for all the differences involved, including the use of watery ink in the first place, the use of a thick vertical plant-like brush to apply it and the pose, in the differences in the physical activities that attracted particular attention in the two regions. If we apply that approach we can say that it looks as if the European artist adopted his pose with brush and palette from the knight with his sword and shield, while the Chinese artist adopted his very different pose from the farmer planting rice in an irrigated field. It is not unhelpful that the artist shown in this print, who could have been shown in the act of representing any subject, is in fact painting a knight, while the base of the Chinese character for painting is in that for field (Fig. 4.7, right). It seems that just as European painters empathized with knights so Chinese painters empathized with rice farmers. In both the empathy and its consequences are unconscious and rooted in neural mechanisms of which those who experienced it would have been unaware. It might be thought that what we are observing is only an imitation of movement, but this is to underestimate what is going on. The interest in knights and rice farmers reflects complex mental engagements. These include social aspiration and material advancement in the case of the European artist and the knight, and in the case of the Chinese artist and the rice farmer physical well-being and prosperity. There are also important associations in terms of mood. It would be impossible to identify with a knight without relishing the deadly killing power of the sword and the defensive effectiveness of the shield, while empathizing with a rice farmer would necessarily involve more gentle and nurturing dispositions. One of the greatest benefits of a neural approach is that it makes it impossible to separate the motor either from the sensory or the emotional.

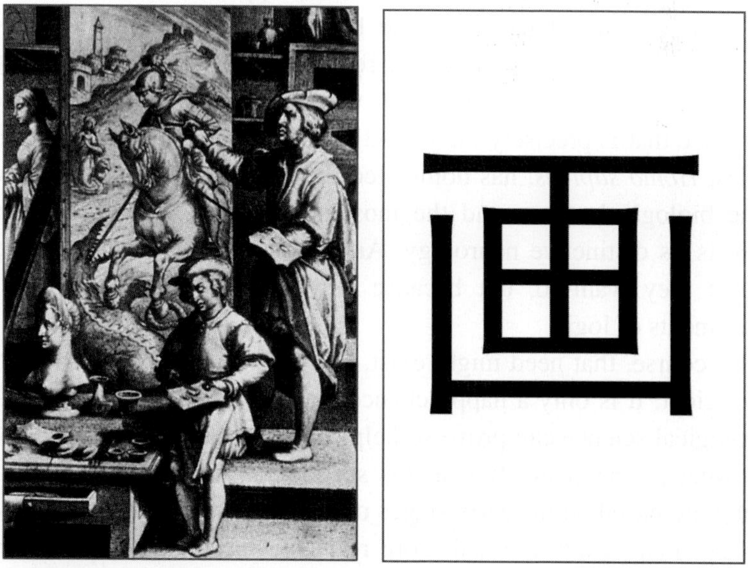

Fig. 4.7. *Left*: European artist painting, engraving. *Right*: Chinese character for "hua" (modern simplified version), painting.

4.5 Conclusion

This discussion of the fruitful impact of imitation on the creative artist also brings us back to our underlying argument, that the art historian can learn from, that is, in effect, imitate, the scientist. One way of presenting what I have been doing is to say that I have been imitating the ethologist. The processes involved in the relationships I have suggested between the activities of artists and those of other members of society are not the same as those that lead some birds to imitate the sounds of chain saws and other machines, but their foundations lie in the same area. In humans, as in birds, the foundations for what are in effect new forms of artistic expression are neural. The art historian can evidently learn much from the ethologist. Art can indeed be better understood using the tools and assumptions of the scientist.

If the argument advanced above is accepted, even the most complex of human behaviors, such as art, can be usefully brought within the domain of science, and this scientific approach to art, far from reducing it to a cold mechanistic activity, brings to it a new warmth and gives it a new life. Nor is this adoption of a scientific approach by an art historian driven, as so often happens in the humanities, by fashion or passing fancy. Rather it follows from a new recognition of fundamental realities, that art is a worldwide phenomenon with a forty thousand year history. That since that is precisely the period during which the particular human species, *Homo sapiens*, has dominated the planet art must have its roots in the biological nature and the most relevant aspect of that species' nature is its distinctive neurology. Art historians do not turn to science because they want to, but because they are forced to do so by the requirements of logic.

Of course, that need might exist, but the neurological science could be deficient. It is only a happy chance that, because of recent advances, neurological science can prove so helpful. Not that we should be content with what science can offer at this stage. The arguments made in this chapter are based on the early stages of a research that is becoming every day both more extensive and more refined. One of the most attractive consequences of adopting a neuroscientific approach to art is that it promises ever richer perspectives on the activity. It is as if one is getting on a train that is only now leaving the station and traveling faster and faster. Another attraction of adopting a neuroscientific approach to art is that one is basing oneself in a field that is constantly being subjected to rigorous testing. While most approaches in the humanities are based on little more than apercu reinforced by analytical reflection, neuroscience is based on experiments, and while approaches based on apercu reinforced by analytical reflection are constantly susceptible to undermining by new data and new observations, the speed and extent to which this happens is largely determined by other factors, such as devotion to intellectual heroes and fear of the new.

In the case of science not only do such factors, although present, have less impact, but more importantly the best knowledge is seen as a common good, its quality dependent on its being constantly subject to testing by experiment and other forms of review. To which it can be added that in the case of neuroscience there is a particular urgency to such testing and review, since the principal application of knowledge is in the improvement of medical care. While a hypothesis in the humanities is likely to be rejected because the framework in which it is presented for some reason no longer seems attractive, a neuroscientific hypothesis will be rejected simply because it can be shown not to work as well as a new one, regardless of its attractiveness. Any of the hypotheses on which I have relied here may be replaced by others in the years to come. That will not invalidate the whole approach, only require its revision. That is why it is so beneficial to base a humanistic activity, such as art history, in a science. Art history that is sustained by neuroarthistory is much surer of renewal than art history on its own.

References

Boardman, J. [2006] *The World of Ancient Art* (Thames & Hudson, London).

Gombrich, E. H. [1993] *A Lifelong Interest: Conversations on Art and Science with Didier Eribon* (Phaidon, London) p. 133.

Holly, M. A. & Moxey, K. (eds.) [2002] *Art History, Aesthetics, Visual Studies* (Yale University Press, London).

Huttenlocher, P. R. [2002] *Neural Plasticity: The Effects of Environment on the Development of the Cerebral Cortex* (Harvard University Press, Cambridge, MA).

Lam, L. [2011] "Arts: A Science Matter," in *Arts: A Science Matter*, eds. Lam, L. & Burguete, M. (World Scientific, Singapore).
Lavin, I. (ed.) [1994] *Meaning in the Visual Arts: Views from the Outside. Centennial Commemoration of Erwin Panofsky (1892-1968)* (Princeton University Press, Princeton) p. 208.
Neidich, W. [2003] *Blow-Up: Photography, Cinema and the Brain* (Distributed Art Publishers, New York).
Onians, J. [1996] "World art studies and the need for a new natural history of art," *Art Bulletin*, June, 206-209.
Rizzolatti, G. et al [1996] "Premotor cortex and the recognition of motor actions," *Cognitive Brain Research* **3**, 131-141.

5

Science and Art in China

Bing Liu

Although the idea of Science and Art (S&A) has been suggested for quite a long time, the real concern for this topic in China was raised only in the last 15 years or so. Since then, many research papers and dissertations have been published in this field. In this chapter, the brief history and status quo of the studies on S&A in China are reviewed, including, e.g., the important events, conferences, exhibitions, and publications. Works on S&A from China are classified into four categories with different depth levels. Existing problems in the development of S&A in China are summarized and discussed.

5.1 Introduction

Science and Art (S&A for short; or, as referred by others, Art and Science) is a rather vague and ill-defined area. It includes many contents and research works that are directly or indirectly related to S&A. It also covers some relevant activities of art creation and scientific research, and even some social activities on this theme. Here, the word science can be broadly understood to include technology.

This chapter provides a brief overview of the development of S&A in mainland China (early period in Section 5.2; later period since 1987 in Sections 5.3 and 5.4). In particular, Section 5.3 covers important conferences and exhibitions, while publications—books, journals, and research papers—on S&A are summarized in Section 5.4, wherein, research works are classified into four categories. Existing problems in the development of S&A in China are summarized and discussed in Section 5.5. Lastly, dissertations on S&A are listed in Appendix 5.1.

5.2 The Concept and Early History

Around the world it seems difficult to pinpoint the time, site, and the person for first clearly putting forward the concept of S&A, linking the two different areas together. In China, it is also the case. Of course, it could happen that this concept simply evolves from the vague to the relatively clear, without a clear-cut point of birth. Moreover, to introduce the concept is one thing, to have a contemporary understanding of the concept of S&A, and to spread it, is another matter. In the latter sense, as far as we know, the relatively early person in China who held such thinking is the well-known educator and scholar, Cai Yuan-Pei (1868-1940) (Fig. 5.1). As early as more than 60 years ago, he proposed that S&A are the "two wings" of human culture—an idea associated particularly with his conviction on education. He said, "All moral persons, without exception, depend on art and science, as a cart with two wheels and a bird with two wings."[1]

Fig. 5.1. Cai Yuan-Pei.

Furthermore, he also pointed out that science belongs to concepts but emotions, art; between them there is interpenetration, such as "acoustics and music; optics and picture; mineralogy and pretty crystals; plant branches and leaves, and feathers; animals and posture beauty"; the two elements in each pair are correlated. In other words, S&A are

[1] All quotes of Cai Yuan-Pei are from [Gong, 1998].

correlated and mutually penetrated. "Those who only study science tend to have a mechanical worldview," he went on; in addition, "They not only have no interest in themselves and no passion to the society, but also show no creative spirit, just copying their predecessors' model, in their own scientific research." Considering the early year these ideas—very similar to our understanding today—were advanced, they were really incredible insights from a true pioneer.

Despite such early pioneering ideas, S&A research in China was neither immediately nor vigorously carried out. Sporadic works appeared, but S&A was not treated seriously as a real enterprise until 20 years ago.

Since S&A is interdisciplinary in nature, it is not part of any traditional subject. Thus, the early works came from some interdisciplinary fields, like philosophy of science and history of science. In the 1950s and 1960s S&A was discussed with a Chinese characteristic, under the so-called "Dialectics of Nature".

It was in this context that in 1985, the book *Fragmental Thoughts on Science, Art and Philosophy* by Zhao Xin-Shan [1985], a scholar in philosophy of science, appeared to be a landmark. Although it does include a lot of discussion on the relationship between S&A, this book, basically, carries a style of essay writing; it is not an in-depth research book. Yet, the appearance of such a publication on S&A signified that some scholars had begun to notice this field.

Hereafter in 1987, a research book *History of Ideas on Aesthetics of Science* [Xu, 1987] appeared (Fig. 5.2), which apparently could be regarded as the beginning of thematic studies in this field. The book, a historical narrative, was written by a scholar in the field of Science of Science. The author attempted to found the new discipline of Aesthetics of Science, using an interdisciplinary approach and taking advantage of the hot topic of "New Disciplines" at that time. The scope of this book is wide, involving ancient and modern times as well as China and abroad. This over-800-pages tome looks like a history-of-science book written from the perspective of aesthetics of science. In the author's own words, aesthetics of science "is a new, comprehensive discipline formed from the combination of natural science and aesthetics; it can be simultaneously a branch of theoretical 'science of science' and a branch of aesthetics." His guiding principle is "to be instructed by Marxism, writing from the epistemological point of view, to analyze the promoting role of aesthetics of science in the history of science, and to analyze the aesthetical value of some important achievements in and theories of natural science." Obviously, such work bears the imprint of that era. For instance, the emphasis on Marxism and the motivation "to creation our own" new discipline, relate closely to the research tradition in science of science in China. Historically, this in-depth study of S&A, using aesthetics of science as the Trojan horse, was groundbreaking and resulted in no small impact at that time.

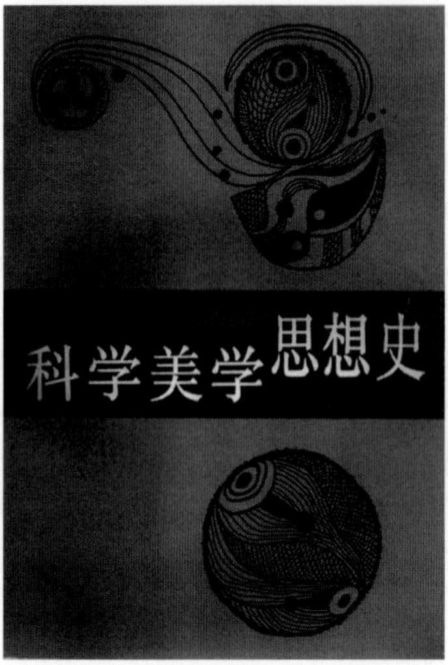

Fig. 5.2. Book cover of *History of Ideas on Aesthetics of Science* (1987).

5.3 The New Rising: Important Symposiums and Exhibitions

Here, some important symposiums and exhibitions on S&A held in China after 1986 are described.

5.3.1 *Symposiums Organized by Tsung-Dao Lee (Since 1987)*

In the last decade or so, the concept of S&A aroused more attention and became the focus in China, due to the contribution of Tsung-Dao Lee, a Nobel laureate and physics professor at Columbia University.

Fig. 5.3. Tsung-Dao Lee.

In 1986 T. D. Lee and the Chinese Academy of Sciences jointly set up the China Center of Advanced Science and Technology (CCAST). In the same year, the first international conference of CCAST was held. Since 1987 CCAST organized annually international conferences, attended not just by scientists but also some painters who were invited to create paintings according to the 22 scientific themes (mainly particle physics, and high-energy physics and cosmology) of those conferences. A tradition was formed.

Moreover, in 1993 and 1995 CCAST also organized S&A symposiums and art exhibitions, respectively, co-sponsored by Yanhuang Art Gallery, and Science and Technology Daily Agency. Because of the prestige of the organizer, those symposiums and art exhibitions were visited by some national leaders; e.g., the Head of State and Chairman of the Chinese Communist Party, Jiang Ze-Min, in 1993. During these symposiums, artists and scientists got together to express their views on the relationship between art and science, to inspire each other. Similarly, these activities received extensive coverage in the media, making the topic of S&A widely known to the common people. The China Stamp Company even issued commemorative envelopes for the S&A Symposium of 1993 (Fig. 5.4).

Fig. 5.4. Commemorative envelope for the S&A Symposium of 1993.

These specially created paintings on specific science topics of the symposium were collected and added with interpretive text about their relationship to the related scientific problems were published as the book *Science and Art* in 2000. In this finely printed book of color pictures (Fig. 5.5) and texts, Lee [2000], the editor and contributor of those texts, put forward his now famous metaphor, called the "fundamental theme" by him: "Science and Art are as inseparable as the two sides of the same coin." And, "They share the common ground that is human creativity. They both aim at a universality that represents truth." Later, this "coin metaphor" became the most-cited visualized expression by Chinese academia and non-academia alike when addressing the issue of S&A.

Fig. 5.5. A painting from the book *Science and Art* [Lee, 2000].

In October 2000, the launching ceremony and symposium organized for the book *Science and Art* was held in Yanhuang Art Gallery, Beijing. I was there by invitation. I observed that although many scientists and artists did attend and make speeches, no scholars in the field of S&A were present; moreover, the speeches were based on personal impressions rather than on research results.

5.3.2 *International Conferences and Exhibitions Organized by Tsinghua University (2001 and 2006)*

In May-June, 2001, Tsinghua University in Beijing hosted the most grand-scale International Symposium of Art and Science (A&S) and the accompanied International Exhibition of A&S. Artists and scientists from 19 countries/regions submitted more than 200 research papers and 566 exhibits—154 exhibits from abroad and 412 from China (Fig. 5.6). The exhibits filled up 10 exhibition halls in the National Art Museum of China. They could be divided into two groups: *artworks* representing the mutual understanding between A&S, and *art designs* reflecting the combination of A&S. The submitted papers focused on four themes: (1) A&S was originally together as one, (2) science calls for art, (3) art in the light of science, and (4) combining A&S will benefit humankind [Li & Zhou, 2001]. This symposium and exhibition have promoted tremendously the importance of A&S throughout China.

Fig. 5.6. A sculpture presented at the first International A&S Exhibition (2001) organized by Tsinghua University, Beijing (now in campus after relocation).

In October 2006, Tsinghua University hosted the second International Exhibition and Symposium of Art and Science (Fig. 5.7). This time, participants came from more than 50 institutions in 21 countries/regions; submitted over 120 papers and more than 200 exhibits, including works on computer-interactive art, virtual-reality art, animation art, Web art, BioArt as well as art designs and fine-art works combining A&S.[2]

Fig. 5.7. The second International Exhibition and Symposium on A&S (2006) organized by Tsinghua University, Beijing.

5.3.3 *The Science, Art (Aesthetics) and Innovation Forum (2007)*

In December 2007, Chinese Society of Dialectics of Nature, Southwest University, and some other institutions in China organized the forum on Science, Art (Aesthetic) and Innovation: Interdisciplinary Theory and Research Methods, at Southwest University, Chongqing (Fig. 5.8). That "innovation" appeared in the forum title reflected the trend and emphasize in China at that time. In reality, the forum's focus was on promoting interdisciplinary studies integrating humanities/social sciences

[2] "The Second International Exhibition and Symposium on Art and Science opened in Beijing," *Sculpture*, No.6, 80 (2006).

with natural science. That "aesthetics" appeared in the title was due to one of the organizers, Zhao Ling-Li of the Institute of Higher Education, Southwest University; Zhao once worked experimentally in psychological study of aesthetics. Of course, the use of experimental psychology method to study aesthetics (borrowed from natural science) was also a new trend in the field of S&A.

During the forum, some young and middle-aged scholars from various Chinese universities, colleges, and institutes participated in the discussion. Another novelty is that a number of vivid live debates among scholars took place in the university campus, with the participation of many undergraduate and graduate students. Some papers from the forum were selected and published in the first issue of the journal *Art Education in China* (2008).

Fig. 5.8. The Science, Art (Aesthetics) and Innovation Forum, Chongqing (2007).

5.3.4 *The Shanghai International Science and Art Exhibition (Since 2004)*

Starting 2004, an annual Science and Art Exhibition[3] had been held in Shanghai, which became the largest and longest series of S&A activities in China (Fig. 5.9). During the exhibitions, works by artists from around the world were interspersed with some thematic exhibitions; seminars,

[3] Renamed International Science and Art Exhibition in 2005.

discussions and lectures addressed to the public were also organized. The reason that these exhibitions are able to be continued serially in a large scale is that the Shanghai government and the Shanghai Science and Technology Association want to provide distinctive popular activities for the public. Actually, these exhibitions can be considered science-popularization (or science communication) activities, fulfilling Shanghai government's obligation to implement the "Project of the National Action Plan for Scientific Literacy" set down by the central government [Chen & Wang, 2008].

Events like these suggest an important growth direction for S&A in China, i.e., combining the study and activities of S&A with the work of mass communication, especially science communication.

Fig. 5.9. The Shanghai International Science and Art Exhibition (2007).

5.3.5 *The Beijing International Conference on Science and Arts (2010)*

Something different occurred in the International Conference on Science & Arts, July 10-11, 2010, Beijing (Fig. 5.10), which has "Arts" in the title, for the first time in China. The conference was organized by the College of Humanities and Social Sciences (Graduate University of Chinese Academy of Sciences) and Beijing Association for Digital Science and Technology Popularization, managed by Li Da-Guang and Liu Ying, and held at the Graduate University.

Fig. 5.10. The International Conference on Science & Arts, Beijing (2010). The names of 10 invited speakers are listed on the right-hand side of the two posters; the photograph exhibition and a "Science & Arts Evening," the left-hand side of the lower poster.

The conference listed 48 abstracts and had 34 talks,[4] with over 180 participants. There was also an exhibition of photographs taken in China and France by Charles Chamberoque during the early 1980s. A very interesting "Science & Arts Night" was held the first day, which included performances of classical and multimedia dances, popular-science comic monologue, and animation short films from Canada. The conference proceedings will be published by Tsinghua University Press. The conference was a success; a new series, starting with this one, is under planning.[5]

5.4 Publications on Science and Art

For the development of a research field, both academic and popular publications could play an important role. Compared to foreign countries, S&A in China is still lagging behind. Thus, translation of foreign works plays a very prominent role. Although translation works in the field of S&A are still rare, their influence on the development of this field cannot be ignored.

5.4.1 *Books*

In view of the fuzzy demarcation of S&A as a research field, many translated books on art theory, art criticism, design theory, philosophy of science, and history of science can be included in this category. But here only a few directly-related books are introduced.

Foremost, we should mention a very peculiar book, Douglas Hofstadter's *Gödel, Escher, Bach: An Eternal Golden Braid* [1980], which won a Pulitzer Prize. It discusses many topics (mainly mathematical logic, computer science and artificial intelligence) and

[4] Among these, there were 10 invited talks of 30 min. each, given by speakers from Canada (John Blouin, Martin Legault), China (Peng Pei-Gen), France (Ghislaine Azémard, Jean-Louis Boissier, Charles Camberoque), Germany (Hans Dehlinger), Japan (Yoichiro Kawaguchi), Switzerland (Daniel Pinkas) and USA (Lui Lam).

[5] After this Beijing conference, the First National Forum on Science, Art and Philosophy was held in Xichang, Sichuan Province, Aug. 16-18, 2010, sponsored by Chinese Society of Dialectics of Nature and Sichuan Association of Science and Technology.

covers some background on music and painting. As stated by the author, the way of narrative in this book is unorthodox. The book is even too difficult for ordinary readers. However, in China, as early as 1984, it was rewritten and simplified into a little book with the same title [Yue, 1984] (Fig. 5.11, left), published in a very popular book series called "Going to Future." Then, for quite a long time this little book impressed many people, inducing them to think about the relationship between S&A. In 1996, the complete and very fine translation of Hofstadter's book was published by the Commercial Press [Guo et al, 1996] (Fig. 5.11, right). In contrast to the simplified version, relative few readers find this complete version understandable; most readers simply shelf it for collection.

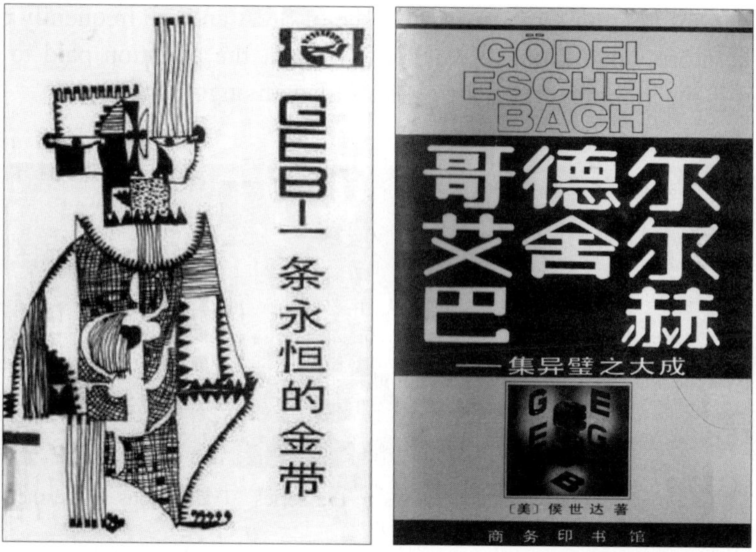

Fig. 5.11. Two Chinese translations of Douglas Hofstadter's *Gödel, Escher, Bach: An Eternal Golden Braid* [1980]. *Left*: Simplified version [Yue, 1984]. *Right*: Complete version [Guo et al, 1996].

It should be noted that a few popular science books at the high end could be counted as relevant to S&A. For example, among the famous "The First Cause" book series published by Hunan Science and Technology Press, books translated from *The Quark and the Jaguar:*

Adventures in the Simple and the Complex (M. Gell-Mann) and *Truth and Beauty* (S. Chandrasekhar) belong to this class, which are frequently cited in research works related to S&A.

In 2000-2003, I edited a set of books on S&A, called "Translations in Aesthetics of Science," published by Jilin People's Publishing House. There are five books in the first series (Fig. 5.12), Chinese translations of some classics: *The Curves of life* (T. A. Cook), *Art & Physics: Parallel Visions in Space, Time, and Light* (L. Shlain), *Beauty & Revolution in Science* (J. W. McAllister), *Emblems of Mind: The Inner Life of Music and Mathematics* (E. Rothstein), and *The Music of the Spheres: Music, Science, and the Natural Order of the Universe* (J. James). The central theme of this series is the beauty of nature and the beauty of science. As far as I know, these books are still a relatively complete set of Chinese translations bearing directly on the issue of S&A and are frequently cited. Furthermore, apart from researchers in S&A, the attention paid to this series is much more among art workers than among scientists.

Fig. 5.12. Front covers of the five books in the series "Translations in Aesthetics of Science."

In another subfield of S&A, Aesthetics of Environment, there exists a set of books called "Environmental Aesthetics Translations," edited by Arnold Berliant and Chen Wang-Heng, and published by Hunan Science and Technology Press in 2006. The books translated include *Nature and Landscape* (A. Carlson), *The Crazannes Quarries* (B. Lassus), *The Aesthetics of Environment* (A. Berleant), and *The Beauty of Environment* (Y. Sepänmaa).

Finally, two more books could be mentioned. One is *Science and Art* written by Dai Wu-San and myself [2006]. Another is *Readings in Art and Science* [Dai & Liu, 2008], a collection of articles selected from the Chinese research literature on S&A; the book is designed with teaching in mind.

5.4.2 *Journals*

Quite a number of Chinese articles on S&A have appeared in journals or popular publications. However, these entire journals are in the field of art, science, social sciences or humanities, with the exception of one. The one exception is the journal *Art & Science*, devoted exclusively to S&A, which is edited and published by Tsinghua University in Beijing (Fig. 5.13).

Fig. 5.13. Selected issues of *Art & Science*, a journal published by Tsinghua Univiersity.

Starting from 2005 this journal has published nine issues till 2009; most research papers are on A&S. In this Journal, the contents are grouped under the sections: Thought, Vision, Explore, History, Panorama, Field Work, Translation, and Book Review. In addition to Chinese authors, many scholars from various countries/regions also contribute (while all papers are published in Chinese). Starting from the second issue, in addition to regular papers, every issue has papers addressing one or two special themes. These themes are: ancient Chinese thoughts of mechanical design and technological design, ancient Chinese images, art and exploration by psychological, physiological and anthropological approaches, art and archaeological research, art and geographic research, modern design, art and four seasons, art and sports, local knowledge, and Buddhist art, etc. However, the publishing time of this Journal is still not regular enough and quite a number of papers are on traditional study of art, not related closely to science.

5.4.3 *Research Papers*

Apart from popular articles published in newspapers, a recent Web search through the Chinese Journal Retrieval System in CNKI produces 285 articles using the key words "science and art" and 178 articles using "art and science." Even though this is not a very thorough search, it does reflect the state of research in this area within the last decade in China. Note that these papers include those that are related to but not strictly on S&A studies.

It is noteworthy in China that the number of master and doctoral theses on S&A starts increasing in recent years. Twenty theses are found[6] within 2000-2008: 19 master theses and 1 doctoral thesis. See Appendix 5.1 for more details.

The research on S&A in China can be roughly divided into four categories, corresponding to four depth levels.

1. *First category*: Works (done at the early stage) viewing the relationship between S&A superficially. For instance, these works focus on those established scientists who are cultivated in art or

---

[6] These dissertations contain the phrase "art and science" or "science and art" in their Key Words, and are thus searchable in the Web.

point out superficial the similarity between A&S [Gong, 1994; Gong, 1996].

2. *Second category*: Works that discuss correlation between S&A at the technical or technological level. Examples are those that focus on how technological means or new materials are applied in creating art, or helping to express art in new forms; e.g., digital art and multimedia art [Zhu et al, 2004; Zhang & Wei, 2006]. These works are practical in nature. A large part of the art-design studies belongs to this category.
3. *Third category*: Works that discuss cognitive issues and concepts in A&S, focusing on the mutual influence, mutual inspiration and mutual penetration between the two. Examples include those studying the role of aesthetics in the process of scientific research, and the impact of scientific ideas on art creation [Yan, 1992; Yao, 1995]. These works represent the understanding of A&S at a higher level, making it possible to find new expressions and new discoveries in either art or science. Most of such works are related to studies in philosophy of science or art theory.
4. *Fourth category*: Works that consider S&A from epistemological and ontological perspectives, at a philosophical and fundamental level. It may bring about new insights on the essence of humans' understanding of nature, and the essence of nature (with humans included) itself [Gao & Yang, 2002; Liu, 2004]. Unfortunately, such studies are still rare.

5.5 Problems in Developing Science and Art in China

Presently, S&A research in China has improved considerably but is still at its initial stage of development. The major problems remaining are as follows.

1. Due to inadequate academic accumulation, there is very few high-level research work; most are casual arguments, not in-depth studies. Comparing to other research fields, the amount of research on S&A is much less.

2. Very few researchers are specialists in S&A studies. They engage in it because they are interested in the topic but are doing it part time. Moreover, artists and scientists involved in these researches are limited by their training in science and art, respectively, leading to obstacles in combining S&A.
3. In some very influential S&A exhibitions, taking that in Shanghai and the two hosted by Tsinghua University as example (see Sections 5.3.2 and 5.3.4), many of the exhibits are barely related to S&A. This is still a common problem in such kind of exhibitions.
4. The institutionalization of S&A is not yet completely developed. For example, there are still no academic institutions or professional organizations as well as formal journals on S&A.[7]
5. As a consequence of institutional constraints, S&A research and education are still in a very difficult situation; especially in the universities or colleges, due mainly to the official setup and managing system of the disciplines.[8] Presently, S&A is *not* on the official list of allowed disciplines. This managing system makes it difficult for S&A research to be acknowledged by the academic system, and greatly restricted the possibility for S&A to get research funding.

On the other hand, there are some favorable conditions for the development of S&A as a research field.

1. Some prominent individuals actively advocate and participate in works related to S&A.
2. Government support for science popularization makes it possible for some S&A works to go ahead in the name of science communication.

[7] Strictly speaking, Tsinghua University's *Art & Science* is only a book series, rather than a formal journal with an appropriate ISSN number.

[8] In China, setting up a new discipline in a university requires the approval of the Department of Education, which does not approve it case by case but revises the official discipline list infrequently. Research works have to be put, one way or other, into an approved discipline: art, art design, subject teaching, teaching and training in physical education, aesthetics, Chinese philology, education management, foreign art history, early childhood education, philosophy of science and technology, etc.

3. Education reform makes the education field a very important area to introduce S&A studies.
4. Society's passion for science, technology and modernization, and the country's rapid economic development help research works related to art design in getting people's attention.

In conclusion, when everything is considered and looking from the long-term perspective, S&A remains an exciting and very promising field in China.

Appendix 5.1: Science-and-Art Theses in China

Master and doctoral theses on S&A from China since the year 2000 are listed in Table 5.1 here.

Table 5.1. Master and Ph.D. theses on S&A since 2000. Thesis No. 8 is a Ph.D. thesis; the rest, Master theses.

No.	Title	Author	University	Year
1	Differences between Styles of Writing in Science and Art	Gao Ming-Yang	Yunnan Normal Univ.	2000
2	Integration and Development of Art and Science Factors in Landscape Design	Quan Jin	Wuhan Univ. of Technology	2002
3	Creative Thinking and Scientific Method in Art Design	He Li-Hua	Wuhan Univ. of Technology	2002
4	Science and Art: Homological and Interactive	Wang Lei	Guangzhou Univ. of Chinese Medicine	2002
5	Integration of Art Education for Young Children and Science Education	Liu Zhi-Hong	Nanjing Normal Univ.	2003
6	Jin Dynasty-Style Bronze from the Perspective of Science and Art	You Zhitao	Shanxi Univ.	2004
7	Coordinated Development of Arts and Science	Yang Yan-Hong	Hebei Univ.	2004
8	Scientific Talent Identification and the Application System of Chinese Elite Athletes in Rhythmic Gymnastics	Guo Xiu-Wen	Beijing Univ. of Physical Education	2005
9	Relationships between Beauty in Science and Beauty in Art	Li Sheng-Nan	Jilin University	2006

10	Experimental Research on Integrating Infant Science and Art Enlightenment	Huang Hai-Tao	Shandong Normal Univ.	2006
11	Gombrich and Popper	Liu Guo-Zhu	Nanjing Normal Univ.	2006
12	Beauty in Science and Beauty in Art in Panorama	Wang Li-Xin	Capital Normal Univ.	2006
13	Beauty in Science and Beauty in Art in Art Design	Zhang Shi-Wen	Northeast Normal Univ.	2007
14	Comprehensive Teaching that Integrates Art and Science	Liu Hui	Inner Mongolia Normal Univ.	2007
15	Harmony and Unity between Rationality of Science and Geist of Art	Meng Qing-Lin	Nanjing Arts Institute	2007
16	Integration of Science and Art	Wang Xuan	Henan Univ.	2008
17	Scientific Thinking Promotes Artistic Development: Theory, History and Model	Chen Guang-Lin	Guangxi Normal Univ.	2008
18	Science in Painting: Leonardo da Vinci's Art and Science	Wang Xin-Xia	Shandong Univ.	2008
19	Differences of the Modifier-Noun Phrase between the Artistic Style and the Scientific Style	Yang Xu	Jinan Univ.	2008
20	Differences of the Adverbial-Verb Phrase between the Artistic Style and the Scientific Style	Cheng Heng	Jinan Univ.	2008

References

Chen, Ji-Fang & Wang, Xiao-Ping [2008] "Shanghai Science and Art Exhibition," *Art Education in China*, No1, 117-119.

Dai, Wu-San & Liu, Bing [2006] *Science and Art* (Tsinghua University Press, Beijing).

Dai, Wu-San & Liu, Bing (eds.) [2008] *Readings in Science and Art* (Shanghai Jiaotong University Press, Shanghai).

Gao, Cha & Yang, Xiao Ming [2002] "Science in arts: A possible direction in philosophy of science and technology," *Journal of Zhanjiang Teachers College*, No.5, 14-17

Gong, Cheng-Guo [1996] "Scientists and piano," *Piano Artistry*, No. 5, 63.

Gong, Zhen-Xiong [1994] "The common rules in science and art," *Art & Design*, No. 1, 8-9.

Gong, Zhen-Xiong [1998] "Cai Yuan-Pei: Science and art," *Democracy & Science*, No.3, 36-37.

Guo, Wei-De et al (transl.) [1996] *Gödel, Escher, Bach: A Collection of Different Gems* (Commercial Press, Beijing).

Hofstadter, D. R. [1980] *Gödel, Escher, Bach: An Eternal Golden Braid* (Random House, New York).

Lee, T. D. (ed.) [2000] *Science and Art* (Shanghai Scientific and Technical Publishers, Shanghai) p. 148.

Li, Sha & Zhou, Zhao-Kan [2001] "Art and science dialogue: Sidelights on the *Art and Science International Exhibition* and the *International Symposium*," *Art*, No.6, 52-54.

Liu, Bing [2004] "The parallel between physics and art in knowing the world: In the example of Einstein and Picasso," *Science* (Shanghai), No.6, 42-45.

Zhang, Hong Chun & Wei, Xiao Juan [2006] "On the influence on development of art by progress in science," *Journal of Guangxi University for Nationalities (Philosophy and Social Science Edition)*, No. S1, 110-112.

Zhao, Xin-Shan [1985] *Fragmental Thoughts on Science, Art and Philosophy* (SDX Joint, Beijing).

Zhu, Hui et al [2004] "Science and art in the information age," *Joural of Hebei Institute of Architectural Science and Technology (Social Science Edition)*, No. 4, 119-121.

Xu, Ji-Min [1987] *History of Ideas on Aesthetics of Science* (Hunan People's Publishing House, Changsha).

Yan, Li [1992] "The epistemological significance of beauty of science," *Science, Technology and Dialectics*, No.2, 25-26.

Yao, Bo [1995] "The blending of science and art from the perspective of development in modern fine art," *Journal of Huaqiao University (Philosophy and Social Science Edition)*, No.1, 47-51.

Yue, Xiu Cheng (transl./ed.) [1984] *GEB: An Eternal Golden Braid* (Sizhuan People's Publishing House, Chengdu).

6

The Development of Science Theater

Ivo Schneider

This chapter on science theater deals with those plays on stage in which science plays an important role. Science theater connects scientific activity and scientific results with the emotional and social realm of human life. Its roots can be seen in the dialogues of Galilei, de Fontenelle or Algarotti in the 17th and 18th century, albeit these dialogues could well have but never did enter the stage. Since the beginning of the 17th century, the spectrum of science theater has ranged from unreserved approval combined with the conviction that an improvement of social and economical conditions can only be achieved by scientific progress to fear and refusal of science. Different from other countries science theater in Germany could never claim any status except for a short period, during which it pointed mainly to the negative aspects of scientific activity. The reason behind this unique phenomenon is explored and explained as due to the persistence of the Bildungsbürgertum mentality in Germany. Presented also in this chapter is a long discussion of the science theaters staged around 2000 in New York and London, those in German speaking countries (*Life of Galileo*, *The Physicists*, and *In the Matter of J. Robert Oppenheimer*), the two most successful science theaters of recent years (*Copenhagen* and *Infinities*) as well as science theaters authored by scientists (e.g., Carl Djerassi) and historians of science (e.g., Marc Friedman). France is discussed as an example for a nearly continuous development of science theater from the 17th century.

6.1 What Is Science Theater?

Science theater has different connotations. In today's science centers some of them claim a science theater. This holds for the *Phaeno* in Wolfsburg or the new science center in Hamburg, which will be finished presumably only in 2011. Both offer places for more than 200 spectators in an auditorium which resembles that of a classical theater. But what the visitors see are not the plays performed by actors and actresses but scientific experiments or, in short, science, the results of which are projected on huge screens. Even this kind of science theater has, at least in Germany, a history dating back to the 19th century when in 1889 the science theater *Urania* was opened for the public in Berlin. At this time instead of projections on a screen huge pictures were painted which served as a backdrop for the texts which were read before the audience.

Independent of the way science is displayed, first of all visually in these science theaters, some of which might be even considered by some as art; this connotation of science theater is not considered here. Rather I shall concern myself with the theater of plays, playwrights, directors, actors and actresses. Classical theater in this sense becomes science theater when science plays an important role in a play. The reasons why and how science gets involved in a play are manifold and differ from country to country; they may depend on social, political or cultural differences. This means also that science theater as part of classical theater flourished at different times in different countries. My own encounter with modern science theater took place in 2000 when I learned that several theaters on and off Broadway staged plays, which dealt with mathematics and physics.

6.2 Science Theater Plays Staged Around 2000 in New York and London

In the 1990s, the already anti-intellectual stance of most Americans was compounded by an increasingly critical attitude toward natural science and technology, which was not the case during the preceding Cold War period. No one in the United States thus seemed prepared to accept to what extent bridges between the two cultures—the humanities and the science—have been built in the theater world few years later. Thus, New York theater critics rubbed their eyes in astonishment when they started seeing an increasing number of plays on Broadway beginning in the late 1990s, all dealing with the worlds of mathematics and physics—the most prominent of them certainly being Michael Frayn's *Copenhagen*. The list of plays dealing with physics includes *Space* by Tina Landau in the Public Theater, which was performed there in late 1999 and deals with cosmic galaxies; *Moving Bodies*, Arthur Giron's biography of the physicist Richard Feynman at the Ensemble Studio Theater where it could be seen in April 2000; and *Now Then Again* by Penny Penniston, which finally ended its extended run at the Ivanhoe Theater in Chicago in May 2001.

Penniston's play deals with the amorous adventures of young physicists whose temporal development up to a certain point in act three and then back again reminded delighted connoisseurs of a diagram named after the physicist Feynman. Similar to Frayn in his play *Copenhagen*, Penniston presupposes in her play knowledge of Heisenberg's uncertainty principle, Bohr's complementary principle or the phenomena accompanying nuclear fission. She commented on the success of such plays, found unexplainable by American theater critics, in the following way:

> They [physicists] are not that different from other people ...They are regular people who happen to be very bright. The difference lies in their being obsessed with science the way the rest of us are obsessed with other things that are important to our lives. I wanted to show how that passion for their work affects them in life, and love.

If one could connect the theater world's interest in physics, reawakened after the Cold War phase, with the developments that followed the Manhattan Project, a similar attempt at explanation fails today with regard to the theater plays that are devoted to mathematics and above all to its purest form, number theory.

Hypatia or the Divine Algebra by Mac Wellmann, at the SoHo Rep and Ridge Theater, told in May and June 2000 the story of the life and work of Hypatia, the woman mathematician killed by Christians in Alexandria in late ancient times, who miraculously reappears every few centuries as the drama progresses and presents the world with a new gift of her imaginative powers. The play certainly owes its existence to feminist inspiration; after all, phenomenal achievements in fields such as mathematics were not exactly considered the preserve of women for many years.

The same direction is taken by a movie made in the United States in 1997 by director Lynn Hershman Leeson, called *Conceiving Ada*. Its heroine is Lady Ada Lovelace, daughter of Lord Byron and the first-ever woman to create a computer program. In the film Emmy, a present-day computer freak, develops a machine that puts her into direct contact across two centuries with Ada Lovelace, enabling her to rescue Ada's mathematical ideas for posterity before they are lost through her premature death.

Very large natural numbers are involved in *The Five Hysterical Girls Theorem* by Rinne Groff at the Target Margin Theater, in which for instance a proof is given that the number 2, to the 67th power, minus 1 is not a prime number. The main figure in the play is the number theorist Moses Vazsonyi who tries to complete his latest work concerning a special class of prime numbers which he likes to call the Hysterical Girls during a conference on number theory organized by colleagues in England of 1911. An even more exciting theme was considered in *Proof* by David Auburn in the Manhattan Theater Club, aimed at understanding the way in which mathematicians work by taking number theorems and their proofs as its examples. Although Auburn consulted with mathematicians from the Courant Institute for his New York stage version, critics had expected more insight into the work of mathematicians than what is provided by the protagonists' strokes of genius, which seem to come all too easily. Nonetheless *Proof* was awarded the 2001 Tony Award and the 2001 Pulitzer Prize for drama.

Representing the world of mathematics as authentically as possible is also the main concern of Gus van Stunt in his movie *Good Will Hunting* from 1997, which at least grasps the emotional dimension which the science of mathematics is usually denied. The movie examines the relationship between a Fields Medal-winning math professor at an elite American university and a young man entrusted to his care who is mathematically brilliant yet also a criminal. Like *Conceiving Ada*, *Good Will Hunting* has its strengths in the dialogues and monologues which could easily make up for a science theater play.

Furthermore, a musical entitled *Fermat's Last Tango*, by Joshua Rosenblum and his wife Joanne Sydney Lessner, with the mathematician Andrew Wiles for its subject, was staged by the New York Theater Company in the end of 2000, in celebration of the proof of the 17th-century mathematician Pierre de Fermat's famous "Last Theorem" by Andrew Wiles in 1995. The proof itself is so complex that most mathematical specialists have so far been unable to judge its correctness. Nevertheless, the publication of the solution excited journalists all over the world and—for a brief period at least—was a matter of great public interest, not least perhaps because the best mathematicians had spent some 350 years on the problem but failed to prove the theorem, which Fermat had once jotted down in the margin of a book.

The list of examples of the new connection between mathematics, physics, theater and film in the United States and Great Britain can be extended. A play by Robin Hawdon, *God and Stephen Hawking*, is about the Cambridge physicist Stephen Hawking and his concept of the origin of time. It was staged in the Oxford Playhouse in 2000 but never made it to London. It is about the conflicting positions of religion and science in an unauthorized biography of Stephen Hawking. The play put the cat among the pigeons in the English press because Stephen Hawking himself took objection to being portrayed on stage after having agreed to it primarily [Durrani, 2000].

When I had learned about these activities first of all in American theaters and found nothing comparable in the German theater world I tried to find reasons for this obvious difference in the attitude towards science theater. In this regard I published three articles [Schneider, 2000a; 2000b; 2001] in leading German newspapers and in the organ of the German Mathematical Society (DMV), which met to a large extent with great approval. Subsequently, many articles dealing with science theater appeared, most of them written by English speaking authors. The most comprehensive presentation of science theater appeared in 2006 as a book with the title *Science on Stage* [Stepherd-Barr, 2006]. The author Kirsten Shepherd-Barr is currently a lecturer in modern drama at the University of Oxford. Her main interest in dealing with science theater, which she considers a paradigm for bridging the gap between the two cultures as lamented by C. P. Snow, is the form of the play and its staging in relation to "theatricality" and "performance". Depending mainly on science plays written in or translated into English, she is like most of the other authors less interested in the factors responsible for the different attitudes towards science theater outside the English speaking world. My own concern here is to describe the situation of science theater in Germany (including Austria and the German speaking part of Switzerland) in contrast to that in other countries.

6.3 Science Theater in German Speaking Countries

Theater in Germany was considered in the last two centuries as a "moralische Anstalt", a moral institution, which should instill the German public, especially the German bourgeois, with the social and political values and virtues considered as desirable by the government and/or some political party. Despite the fact that many German plays seem to contradict the function of a moral institution because they instill nothing but laughter or skepticism against any kind of political activity, it was the expectation that theater serves or should serve as a means to educate the general public, leading to the public support of theaters by the state or the community which lasts until today. (See Fig. 6.1 for an example of perhaps the most prestigious German theater.) Only in the last few years that critics of this public support of German theaters hinted to the fact that the theater has lost its former place to other media like TV which inform the spectator in a more convenient visual way. Even more important is the difference in the number of those informed by theater and TV, respectively. Very successful theater productions reach and that only after some years, an audience of still less than a hundred thousand people whereas a program shown in TV is usually attended by millions.

In Germany science except for a short period in the 20th century (at which we shall look in a moment) never conquered the stage. This has to do with the predilections of the so-called Bildungsbürgertum, the educated citizens who had been trained at least in the 19th and early 20th centuries in Greek and Latin at school and comparatively little in mathematics and the sciences. Until in our times the members of the Bildungsbürgertum despite many changes in the system of German high schools preserved a considerable distance from if not contempt of science, technology and especially mathematics.

Fig. 6.1. Burgtheater in Vienna, the most famous theater in the German speaking world (photographed by Ivo Schneider).

Johann Wolfgang von Goethe, who died in 1832, is still considered as the greatest poet and playwright in German literature. He can still be seen as a personal paradigm for the attitudes and the value system of the German Bildungsbürgertum. Goethe who spent most of his life in Weimar was better informed about the sciences of his time than most of his contemporaries. So he had not only worked in the field of optics but also in chemistry, biology and geology. As an amateur geologist he had accumulated the biggest collection of minerals in Europe.

But different from the scientists in academies and universities who following Newton expressed their results in the language of mathematics, Goethe refused and even condemned mathematics as inapt to describe nature appropriately. Mathematics according to Goethe caused, in contrast to his holistic view, a fragmentation of nature. Goethe considered his *Farbenlehre* (*Theory of Colors*) his most important contribution to science; as such it differs completely from his other works. Only in novels like the *Wahlverwandtschaften* or *Wilhelm Meisters Lehrjahre* he liked to hint to his familiarity with science, here his chemical knowledge, in a metaphorical way. In his dramas, which he had described in a 1785 letter to Charlotte von Stein as the "causa finalis of everything that occurs to and by human beings in the world", science had no place. That Goethe's *Faust* figures in the bibliography for a course “Staging Science” at City University of New York can only be understood as a blatant misunderstanding of its content. Neither Goethe nor the German playwrights of his time noticed the fundamental social and economical changes caused by the industrial revolution.

Goethe had died before the opening of the first railway line in Germany in 1835, but even if he would have witnessed in England the dawn of a new technological era he would not have displaced the horse-drawn carriage by the railway in his plays and novels. One might argue that Goethe, even if his literary production was paradigmatic for the German Bildungsbürgertum, could hardly have influenced the development of German theater until our days. This is certainly true, but also that the successors of the German Bildungsbürgertum of the early nineteenth century share still the same attitude toward science as their predecessors.

As a good old saying has it, Germany is the only country in which a person can gain social prestige by confessing to being poor at mathematics. This lack of acceptance of mathematics in German society, and of the sciences that employ its language and methods, is rooted in the value system of the German Bildungsbürgertum of the early 19th century.

This holds despite the impressive successes of the German chemical and electrical industries in the half century before World War I, accompanied by a euphoric belief in unlimited progress of science and technology. In this period leading German scientists from Hermann von Helmholtz to Max Planck and representatives of German industry like Werner von Siemens requested to acknowledge the domains of science and technology as legitimate and integral parts of an appropriately extended understanding of culture. However, the German Bildungsbürgertum, which different from, e.g., the situation in France, was not forced to prove knowledge in mathematics and the sciences in order to qualify for the typical positions it was aiming at and accordingly refused to adopt such an understanding of culture.

Unfortunately, the value system of the Bildungsbürgertum, largely suppressed during this period of growing belief in progress in the decades before World War I, was once again propagated openly after the defeat of 1918. Oswald Spengler’s speech about the cold hand of rationality is as much a part of this as the drastic cuts in hourly teaching loads for math and physics that were imposed on German high schools by cultural bureaucrats during the years of the Weimar Republic.

After the Nazis came to power, such anti-mathematical attitude was officially and successfully connected with a long-standing anti-Semitism. For instance, because of its abstract mathematical structure Albert Einstein's theory of relativity was defamed as unclear, inconsistent with healthy popular sentiment and common sense, sterile and, therefore, Jewish.

It does not come as a surprise that the only period in which German playwrights of some standing contributed to science theater was after World War II. Despite its great success, it was relatively short-lived and mainly informed by a one-sided negative picture of science. Far from respect and recognition for scientific and technological achievements, the plays of the German science theater expressed first of all fear. It was fear of the immense destructive potential of the nuclear weapons developed and produced during the cold war. The main aim of the German playwrights was to stress the responsibility of physicists for the whole of mankind as their first and foremost obligation. According to the specific situation in Germany it was clear that the playwrights did not dare to expose their audience to any kind of scientific terminology, method or result except for very general statements.

6.3.1 *Brecht's* Life of Galileo

The first of these playwrights is Bertold Brecht who presented the different versions of his plays as the result of a continuous revision. His *Leben des Galilei* (*Life of Galileo*) is not only a good example for this but also his only contribution to science theater. Whereas the first version was written during his exile in Denmark in the late 1930s, the second and third version, which appeared in 1947 and in 1957, a year after Brecht's death, were influenced by the impression of the nuclear bombing of Hiroshima and Nagasaki in 1945. The main differences between the first and the following versions concern Galilei's revocation of the heliocentric view according to which the earth turns around the sun. Galilei who was convinced to have proved the heliocentric system of

Copernicus, was forced by the Pope to revoke his conviction. In the first version the revocation appeared as an artifice in order to continue his research uncontrolled by the church in Rome. In the second version for which he collaborated with Charles Laughton in his American exile, Brecht made Galilei a criminal who did not have the guts to resist the Pope's request and so failed to create an ethos, according to which the abuse of science should be hindered. The second version ends with the self-accusation of Galilei in this sense. In the third and last version Brecht had added a scene in which a former student of Galilei smuggles the manuscript of Galilei's final version of his new science, the *Discorsi*, into the Netherlands. So in the third version Galilei could repair his image of a criminal with which Brecht had left the theatergoer in the second version.

6.3.2 *Dürrenmatt's* The Physicists

In some respect Brecht's *Life of Galileo* paved the way for Friedrich Dürrenmatt's *Die Physiker* (*The Physicists*), which received its premiere in February 1962 in Zürich. With 1598 performances, *The physicists* became in the season 1962/63 the play most performed in German speaking theaters, and was still in the seasons 1982/83 and 1983/84 the most performed play in theaters of the German Federal Republic.

The main characters are three physicists and a lady, a psychiatrist, who directs the private asylum in which the viewer meets the four persons. The reasons why two of the physicists are or became inmates in this asylum are twofold. One pretends to be Newton and the other, Einstein. In reality both are special agents of two powerful states who try to take advantage of the findings of the third physicist, a real genius with the name Möbius. Möbius had found the universal world formula. In order to prevent abuse of his formula he separated from his wife, left his family and finally pretended to be insane which brought him into the asylum. Each physicist had a love affair with a nurse who was sent by the director of the asylum to him. The three nurses were killed by their lovers, as soon as each nurse had succeeded in finding out the real identity of her physicist. The three physicists killed in order to prevent any detection of their real motives and aims to be in the asylum. As long as they were considered as insane they could not be treated like normal criminals. As soon as they appeared sane they would be punished as murderers, so they were forced to stay lifelong in the asylum. So far Möbius seemed to be successful in hiding his results, the papers of which he had burned. But it turned out that the director had secretly copied Möbius's papers before Möbius was able to burn them. So the director of the asylum had manipulated the behavior of her three inmates in a way, forcing them to stay there forever whereas she was free to use or better abuse Möbius' findings.

The climax of the play is reached when the three physicists utter their opinions about the consequences of scientific research in the world of today. This happens when either agent tries to convince Möbius to give his results to the state for which the agent is spying.

The first reminds Möbius of his duty to inform the world of his findings. But he has to admit that he as a scientist has no influence on the politicians who decide what to do with the results of scientific research; all the scientist can do is to vote for a political system in which he can trust. So in the end there is no guarantee that scientific results are not abused. The second tries to bribe Möbius with the Nobel Prize and refuses any responsibility of the scientist for his results.

Möbius himself is convinced that the scientist is not free to decide about the use of his findings. For him the proposals of the two special agents can only lead into catastrophe. Because the risk to exterminate the whole of mankind is too high the only way left to him is to withdraw his findings and so to self-deny his work and existence as a scientist.

Dürrenmatt saw *The Physicists* as paradigmatic for the structure of a drama or in this case of a tragic comedy. The ambiguity inherent in the character of the play was interpreted as foreshadowed by Dürrenmatt's choice of the name Möbius for the creative physicist in the group. The German astronomer and mathematician Möbius is best remembered by the "Möbius strip"—a two-sided and two-edged strip will have only one side and one edge when given a single twist to one end and then both ends conjoined. So the two alternative sides of a drama, tragedy and comedy, become only one or, as the two physicists who are after Möbius put it, are at the same time mad and wise, prisoners and free [Sheperd-

Barr, 2006, p. 80].[1] As a justification for the lack of any physics in the play, Dürrenmatt explicitly stated that it is the only obligation of the playwright to deal with the consequences of scientific (here physical) research but not to confront the audience with the physics behind the results.

6.3.3 *Kipphardt's* In the Matter of J. Robert Oppenheimer

In 1964 Heinar Kipphardt's *In der Sache J. Robert Oppenheimer* (*In the Matter of J. Robert Oppenheimer*) had its premiere. This play belongs to the genre of documentary theater. It is based on the FBI-hearing of Oppenheimer in 1954. Oppenheimer was the leading physicist in the Manhattan project in which the atomic bomb was developed in 1945. He had refused to collaborate with the project of developing a hydrogen bomb. In the McCarthy era he was accused in this FBI-hearing of treason and of disloyal behavior against the USA on the basis of his assumed connections to communism. As a result of the hearing Oppenheimer was deprived of the security clearance necessary to participate in government projects like that of the Atomic Energy Commission (AEC). Only in 1963 he was vindicated by President Kennedy.

Kipphardt used the 3000 pages of the minutes of the hearing as the main source for this play. Since only 1 % or 2 % of this text would have sufficed to fill the time available for a play he had to shorten and condense the hearing considerably. Whereas in the real hearing which lasted about a month 40 witnesses were heard, in Kipphardt's play only six show up, three of them defending Oppenheimer and other three being convinced of his guilt. But even the real statements of these six witnesses would have been much too long for a play. So Kipphardt created a text, which differs in part of the real hearing; hereby he followed the principle "to change as little as possible and as much as necessary". In addition Kipphardt corresponded with Oppenheimer with whom he discussed the

[1] Taken the ignorance of a German audience concerning mathematics, here the Möbius strip, and the fact that Dürrenmatt never referred to the Möbius strip as the reason for his choice of the name Möbius, the otherwise attractive interpretation appears pretty far-fetched.

content of the play. Despite the changes Kipphardt made at Oppenheimer's instigation, Oppenheimer initially threatened to enjoin the performance of the play; later he regretted to have done so.

The main theme in this play about Oppenheimer is the conflict between the scientist who feels obliged to serve with his research activity the whole of mankind on the one hand and the interests of his state on the other. In this sense the character Oppenheimer says that he was more than loyal to the USA when he betrayed the ideals of science by his participation in the Manhattan project. So Kipphardt who first worked as a psychiatrist and later came to the theater as a dramaturge and playwright addresses in his play of Oppenheimer's hearing a conflict similar to the conflict of Brecht's *Galileo* and Dürrenmatt's *Physicists*.

Still in the 1960s, the Austrian Kurt Mellach mainly known as an author of children's books published the play *Archimedes oder der Moment der Physik* (1967). In this play, the famous Syracusan is portrayed as a double agent spying for his home town Syracus and for Rome, like the double agent Fuchs whom the Soviets had smuggled into Los Alamos, the center of the Manhattan Project. However, Mellach's double agent Archimedes was not a success, in contrast to the plays of Brecht, Dürrenmatt and Kipphardt. Perhaps Archimedes with his image as the greatest mathematician and scientist of antiquity was not the right person to convey the idea of a double agent in the 20th century.

6.3.4 *Differences Concerning the Situation of Theaters in Germany and in Other Countries*

One could imagine that the possibility to play with the anxieties of an audience exposed to the outcome of atomic bombing or other ways of destruction designed by scientists could be still a motive for the directors of German theaters to restage a play like Dürrenmatt's *The physicists*. But typically for the current situation in Germany, Dürrenmatt's play was performed in the last years only by a group of 18 girls from a provincial high school in 2000, not in a public theater, but in the assembly hall of their school.

In today's Germany despite all the changes that have taken place, the heritage of the Bildungsbürgertum is still recognizable by its stubborn reluctance to acknowledge mathematics, natural sciences and technology as an independent component of culture. This seems to be the main reason why Germany never developed a tradition of a science theater quite different from what one finds, e.g., in English-speaking countries, in France or in Italy.

In English-speaking countries one does usually not find the prejudices against science and technology like in the German Bildungsbürgertum; in addition theaters receive no public support; they have to secure the financial basis for their existence themselves. The greater German theaters offer with their great companies of actors, part of them tenured, over a season 20 and more different plays, which are performed in the same staging over several years but only once or twice a month. In the English speaking theater world a play is staged for a relatively short period for two to six weeks, say, and every day in the same theater; after this time a new play with a different ensemble is staged. Therefore the success of a play depends on favorable reviews of the premiere and on the positive reactions of the audience in the first few performances. So stage directors and playwrights depend much more than in Germany on public approval of a staged play.

6.4 The Most Successful Plays in the Last Decade: *Copenhagen* and *Infinities*

The different attitudes to science and going with it to science theater in the German speaking world on the one side and in the English speaking world on the other, can best be exemplified by the reception of one of the most successful science theater plays of the last decade: Michael Frayn's *Copenhagen*. The play was a great success in London where it debuted in 1998 and ran for more than 300 performances in the National Theater; and a celebrated triumph on Broadway where it opened at the Royale Theater and ran for 326 performances, awarded with the Tony Award for best play and with the Drama Desk Award for outstanding play in 2000. In contrast, the play in Germany has so far been staged in various small and remote locations and almost went unnoticed. Only the municipal theater in Essen has dared to tackle *Copenhagen*; in 2003 it was staged in the small *Theater 89* in Berlin which specializes in forgotten or in rarely staged plays. Even in France, the staged French version was a great success and received the prestigious Prix Molière for the best play in 1999.

6.4.1 *Frayn's* Copenhagen

In the play, Werner Heisenberg and Niels Bohr together with his wife Margrethe—years after their deaths—try in a fictive meeting to reconstruct the circumstances and motives of a visit paid by Heisenberg to Niels Bohr, which took place in Copenhagen in September 1941. Denmark was occupied by the Nazis at that time which meant that Bohr's house was observed by the Gestapo and that the two men were forced to walk outside in order to speak to one another without any unwelcome witnesses. The walk was short because Niels Bohr had soon become very much irritated and Heisenberg had left the Bohrs shortly after he had returned from the walk. From then on the relationship between Niels Bohr and Heisenberg ceased to be the same as before. For many years neither Heisenberg nor Bohr revealed the content of their conversation. Heisenberg reported about it 15 years later in a letter to Robert Jungk, the author of *Brighter than a Thousand Suns* (1956). Heisenberg claimed in the letter that his aim was to win over Bohr for a common action of all relevant physicists to avoid the development of nuclear weapons, but that he had failed to make himself understood by Bohr. Since Jungk had published an extract from the letter in the Danish edition of his book in 1956 which made it look as if Heisenberg was claiming to have purposely derailed the German bomb project on moral grounds, Bohr felt that this was false and that the 1941 meeting had proven to him that Heisenberg was quite happy with producing nuclear weapons for Germany.[2]

[2] In his letter to Robert Jungk from January 18, 1957 for the new edition of Jungk's book Heisenberg had not claimed to have purposely derailed the German bomb project. Instead he had expressed his conviction that the development of an atomic bomb would be impossible before the end of the war. As long as physicists would work on it politicians and militaries could not decide about its use.

The completely different memories of the two men concerning Heisenberg's visit of Bohr in 1941 opened different interpretations of their conversation. One was that Heisenberg as the leader of the German project to create the so-called Uranmaschine wanted to persuade Bohr to collaborate in this project. Another interpretation is that Heisenberg wanted to know from Bohr about the activities of the allies to create an atomic bomb. A third interpretation has it that Heisenberg informed Bohr about the state of knowledge of those physicists who worked for the Nazi regime in order to warn the allies. In any case, the play deals with a German physicist and a war situation caused by a German regime, reason enough to make it especially interesting and important for a German audience.

According to the publisher of the German version, German theater directors far and wide who said they had a certain interest in the work and requested copies to study, lost whatever interest they had very soon. A reason for such a reaction could be that Frayn's play is, as several critics noted, loaded with scientific dialogue, a little too heavy for the common audience. Though a writer from *Physics World* hailed it as "brilliant theater", a critic from the *Daily Telegraph* wrote, "I felt that my brain was being stretched to breaking point—well beyond breaking point, in fact". So it seems that in Germany the connection between the worlds of mathematics and physics on one side and theater on the other is clearly obstructed by a specifically German mind-set with its tradition — as already mentioned—going back at least to the early nineteenth century.

Copenhagen opens with Margrethe Bohr's question: "Why did he [Heisenberg] come to Copenhagen?" The following drama offers in the first of its two acts different answers discussed between the three persons, to the effect expressed by Heisenberg:

> No one understands my trip to Copenhagen. Time and time again I've explained it. To Bohr himself, and Margrethe. To interrogators and intelligence officers, to journalists and historians. The more I've explained, the deeper the uncertainty has become. Well, I shall be happy to make one more attempt.

Through this sentence shines the physics discussed between the young Heisenberg and Bohr during Heisenberg's first stay in Copenhagen from 1924 to 1926, in which year Heisenberg became professor of physics in Leipzig. The second act deals with the origins of the friendship between the two men and ends with the aftermath of World War II and their relationship in 1947.

The results of Bohr's and Heisenberg's collaboration in the new field of quantum mechanics in Copenhagen was the "Copenhagen Interpretation of Quantum Mechanics" according to which an elementary particle is to be understood at the same time as wave and particle and on Heisenberg's side the so-called "Uncertainty Principle". Presupposing the quantification of the precision of the measurement of a quantity, the Uncertainty Principle claims that the product of the precision of the measurement of two so-called canonical variables like the locality and the speed of an elementary particle in a certain moment is a constant. This is to say if the precision of the measurement, e.g., of the locality of an electron is very high, the precision to the determination of its speed is accordingly low. Applying Heisenberg's Uncertainty Principle to the initial question of the play means that, as the real motives for Heisenberg's motive to come and see Bohr in 1941 is concerned, the concerted efforts of the two Bohrs and Heisenberg to find from more and more details they remember can lead only to a growing uncertainty about these motives. In the discussion between the two physicists who illustrate their findings, e.g., by thought experiments and references to common activities like skiing and table-tennis, Frayn inducts the theatergoer into the main results achieved by Bohr and Heisenberg in quantum mechanics. Then Frayn explicitly and repeatedly transfers the Uncertainty Principle to the human world and to the efforts to reconstruct facts from the memories of the persons involved. So Heisenberg, at the very end of the play, calls the conversation they wanted to reconstruct "some event that will never quite be located or defined" and "that final core of uncertainty at the heart of things". Complementarity is used in the play in a similar way when Heisenberg describes his own status for Bohr in 1941:

> I'm your enemy; I'm also your friend. I'm a danger to mankind; I'm also your guest. I'm a particle; I'm also a wave.

Frayn did not include any stage direction for *Copenhagen*. No gesture, no movement of a character is indicated; there is just the text for the three persons of the play. Nothing distracts the audience from listening to the discussions. Frayn's idea was to have his three protagonists arranged like the components of "an atom intruded by a photon". Accordingly the stage was intentionally bare, excluding three chairs and the three characters. Margrethe, acting as the nucleus, would likely spend most—not all—of the play seated on her chair in the middle of the stage. Bohr was supposed to act as an electron, free to roam about the space. Heisenberg was a photon, an intruder particle; he was only in this atom for the time of his visit.

Copenhagen shows that certain texts, first of all dialogues (even a correspondence understood as a written dialogue), can be staged in a theater as has been done years ago with the correspondence between Voltaire and Frederic the Great in a German theater. The images and metaphors contained in the text of *Copenhagen* and in the staging of it are responsible for the special aesthetics contained in the play. However, it seems difficult for somebody with no previous knowledge of the text and its inherent intentions to appreciate it completely.

Frayn was aware of this and was struck by the totally unexpected success of *Copenhagen*. When he had finished the text he did not dare to offer it to any theater director, convinced that it was without any interest for them; he only hoped that it could be broadcasted.

Copenhagen as staged in London and in New York stirred up heated discussions by historians of science who mainly disagreed with Heisenberg's image given by Frayn. There followed conferences about Heisenberg's role in the Second World War preparing the participants for the play, which was performed as part of such a conference.

6.4.2 *Barrow and Ranconi's* Infinities

The second very successful science theater play, which should be described in greater detail, is *Infinities*. The text of this play by John Barrow was drawn from different mathematical and philosophical sources dealing with infinity. The text for *Infinities* was translated into

Italian. Luca Ronconi, the famous director of the Teatro Piccolo in Milan, staged the play for its premiere in 2002. Ronconi holds that science, here mathematics, affords a new way of presenting theater. As stated in an interview, his argument is that, e.g., in *Copenhagen* the Uncertainty Principle has become part of Heisenberg's biography and so Heisenberg became a real dramatic character comparable to the protagonists in Shakespeare's dramas. This means that the language of the play is the traditional language of drama, which has nothing to do with the discovery of the Uncertainty Principle. From Ronconi's viewpoint [Shepherd-Barr, 2006, p. 200] it was necessary to

> find a meeting point, halfway between drama and science, neither totally on the side of drama nor totally on the side of science, neither something completely "formal" nor a kind of "popularization".

Accordingly the staging of *Infinities*, which consists of five independent parts, does without typical features of classical theater like a plot. The five parts confront the spectators with different concepts of infinity. Since each part requires a separate stage the play could not be performed in the Teatro Piccolo. Instead it was staged in a huge former warehouse in Milan where five different stages were established. The performance of each part lasted about 15 minutes. After a part was finished it was to be repeated immediately but not necessarily in the same form, because the actors as well as their parts could change with every new performance. The spectators could start with any part at any time. After they had seen one part for about 15 minutes they went to another; this meant that the number of possibilities for a spectator to experience the play was infinite.

Infinities deals with the following aspects of the infinite:

1. Hilbert's hotel exemplifies properties of the countable infinite. Hilbert's hotel has countable infinite many rooms and an inter-room phone for all rooms. All rooms of the hotel are occupied. After the arrival of a coach with 40 passengers who want to stay in the hotel the hotel manager has to empty 40 rooms. He calls all the hotel residents and tells everybody to move from his room with number n to room number n

+ 40. By this change of the rooms, room numbers 1 to 40 become available for the newcomers. This is illustrated in the play by a huge monitor which indicates in this and in the following cases who shall leave his room in order to go to another. The hotel manager is than able to provide the necessary number of empty rooms (Fig. 6.2).

Fig. 6.2. The manager of Hilbert's hotel in the 2002 staging of *Infinities* in Milan.

The next coach offers a bigger problem. It transports countable infinitely many passengers with seat numbers 1, 2, 3, ... who want to stay in the hotel. The manager solves even this problem by telling his guests to move from their room with number n to room number $2n$. The passenger of the bus having seat number n is told to take the room with the number $2n - 1$. Here, $n = 1, 2, \ldots$

Perhaps attracted by special features of the hotel or by the cleverness of its hotel manager now arrive countable infinitely many coaches each loaded with countable infinite passengers. The hotel manager can proceed as in the preceding case in order to empty the odd numbered rooms, then puts the first coach's load in rooms 3^n and the second coach's load in rooms 5^n and so on ($n = 1, 2, 3, ...$); for coach number i he uses the rooms p^n where p is the $(i + 1)^{\text{th}}$ prime number.

Another possibility to solve the problem is to look at the license plate numbers on the coaches and the seat numbers for the passengers (if the seats are not numbered, number them). Regard the hotel as coach 0. Interleave the digits of the coach numbers and the seat numbers to get the room numbers for the guests. The guest in room number 1,729 moves to room 1,070,209. The passenger on seat 8,234 of coach 56,719 goes to room 5,068,721,394 of the hotel.

All these measures of the manager of Hilbert's hotel contradict completely the spectator's experience with hotels, which, of course, have only a finite number of rooms. In such a real hotel the number of rooms with odd numbers is obviously smaller than the number of all rooms in the hotel.

In Hilbert's hotel, called "Grand Hotel", the number of rooms with odd room numbers equals the number of all rooms. Here the concept of the *size* of a set comes in. Whereas the size of a finite set is equal to the number of its elements, the size of the subset of rooms with an odd number in Hilbert's hotel equals the size of the set of all rooms. Infinite sets can be defined by the property to have subsets with equal size.

Hilbert's hotel as staged by Ronconi reveals that this new form of the science theater detached from everything biographical can stimulate learning processes as well as emotional reactions.

2. The problems going with immortality. 10 actors and 30 spear-carriers demonstrate that for mortal humans immortality becomes a nightmare of suffering from weariness and boredom. The actors represent very old people sitting and reading apathetically in armchairs or below hair-dryers while at the same time endless monologues suggest the low desirability of continuing such a life forever.

3. Jorge Luis Borges's short story of the Library of Babel leads to the staging of the illusion of a library containing infinitely many books with all possible arrangements of the 26 letters of the alphabet and three punctuation marks, the vast majority of them offering nothing but complete nonsense. But buried among them, somewhere, there are—there must be—books containing every "truth" that anyone could ever want to know. In addition to the infinite library the duplication or

duplicate paradox[3] is enacted by actors wearing the same suits and masks and so create the illusion of identical doubles.

4. Here the dispute between Georg Cantor and Leopold Kronecker concerning Cantor's conviction that the "actually infinite" exists is substantiated by mathematical texts of the two mathematicians, which are written by students of the Milan polytechnic on huge sheets of paper on the walls. The students are sitting intermingled with the audience on tables in a classroom situation.

5. Inspired by Stephen Hawking's reflections concerning time, a time travel is staged by showing a boy in a wheelchair, who aims to run into his grandmother but narrowly misses her hobbling slowly across the stage. Thus the Grandmother Paradox is illustrated according to which it is impossible to return during such a time travel to your grandmother or yourself in an earlier situation. The time travel is also indicated by a train car, in which passengers are sitting looking in both directions.

The reactions to the Ronconi's staging of *Infinities* in Italy were, except for one extremely positive, if not enthusiastic. This holds for both theater critics and representatives of theater studies. The leading Italian newspaper *Corriere della Sera* accompanied and supported the play in the phase of its preparation in different ways. For example, the *Corriere* arranged an exhibition with copies of articles written by leading Italian scientists concerning the interrelationships between philosophy, science and the concept of the infinite published in the *Corriere*, documenting the contributions of the newspaper to throw light on the topic of the play over many years. Over and above, the theater offered an introduction for Italian high-school students planning to attend the play. In the introduction, important concepts were explained and relevant texts were read by an actor of the theater.

In fact the play, with its premiere in March 2002 in Milan and a guest performance in April of the same year in Valencia, was very well received by the audience. It was repeated in Milan in May 2003. All performances were completely sold out despite the relatively high

---

[3] If a person could be materially duplicated two identical individuals would be in two different places at the same time. This contradicts the fact that an individual cannot be in two different places at the same time.

admission price. For the repetitions in May 2003 all tickets were sold two months before the first performance. Already in 2002, *Infinities* won the *Premio Ubo*, the most prestigious prize in Italy for a theater play honoring the best play staged in an Italian theater each year.

There were plans to stage an Arabic version of *Infinities* in the new big library in Alexandria and to find an appropriate place for the staging of an English version. So far none of these plans are realized.

6.5 France as an Example

Here, France is presented as an example for a nearly continuous development of science theater from the 17th century on.

Ranconi's conviction that science theater affords a new form of the theater is shared by the French theater director, Jean François Peyret, who formed his own group of actors, consultants, and coauthors. He staged in 2004/5 *Les Variations Darwin* which was the result of a collaboration with the biochemist Alain Prochiantz; in 2006 *Le cas de Sophie K*, the case of the Russian mathematician Sophia Kovalevskaja; and in 2008 *Tournant autour de Galilée*, commenting Brecht's *Life of Galileo* in the form of a play. All these plays are to a large extent devoid of the personal circumstances under which the scientific ideas of these persons were developed.

Despite his postmodern form, Peyret is in fact a theater representative for a nearly continuous tradition of science theater in France which began perhaps with Molière's comedies criticizing the French ladies who got involved in their salons with the new natural sciences and even carried out experiments on their own. In this regard, even the dialogues of Fontenelle in the 17th century and of the Italian Algarotti in the 18th century can be considered as part of the history leading to French and Italian science theater. It is true that these dialogues were not conceived as plays for two persons, but they could well be performed as such, taking *Copenhagen* as a paradigm. In both Fontenelle's *Entretiens sur la Pluralité des Mondes* and Algarotti's *Newtonianismo per le Dame* (1737), a handsome marquis knowledgeable about the science of his time informs an equally handsome marquess in the erotically loaded atmosphere of a summer night in a park, of the solar system according to Descartes respectively of Newtonian physics. Fontenelle's *Entretiens* were published again and again for nearly a century; they were translated into the main European languages; obviously they met with the expectations of educated female readers.

Although there existed generally in France a greater readiness to consider science and technology as a part of culture and to offer science a place in theater, nonetheless, an author like Voltaire, a contemporary of the old Fontenelle[4] and of Algarotti, and perhaps the best French playwright of the 18th century, never integrated science into his plays. On the other hand, it was Voltaire who had become one of the first French Newtonians. In his *Letters Concerning the English Nation* of 1733 he compared the Cartesian with the Newtonian system of the world albeit Voltaire's main aim in the *Letters* was to show the French public how much better than the French was the English political system. Five years later he convinced himself the superiority of Newtonian science over Cartesianism. He introduced Newtonian science into France with his *Éléments de la Philosophie de Newton*. But different from the case in England where Newton became one of the main topics in English literature of the so-called Augustan age, Newton and Newtonian science never entered into Voltaire's dramas or novels. Perhaps it was this attitude which caused Voltaire to characterize Fontenelle in a very ambiguous manner in his *Micromegas* and to distance himself from Algarotti's *Newtonianismo per le Dame* as not being to the French taste.

The enormously great social changes that went with the industrial revolution had not yet reached France during Voltaire's lifetime. It was five years after Voltaire's death that one of the oldest dreams of mankind, the ability to fly, was realized by the Montgolfier brothers and by Jacques Alexandre Charles. Yet, the event did not earn a single line by the German playwrights at the end of the 18th century. In contrast in France, several plays celebrating this secular event appeared. So in 1784 the play *L'amour Physicien ou l'Origine des Ballons* was staged in Théatre de l'Ambigue Comique in Paris. The hero of this love story is Zélindor, who had to convince the father of his beloved Almenzine of his skills and expertise in science and technology by the invention of the hot

---

[4] Fontenelle died in 1757 missing his 100th birthday by only a few weeks.

air balloon in order to win her hand. Already before and still in 1783 the amateur playwright Bodard du Tezay had celebrated triumphs with his *Le Ballon ou la Physicomanie* in the Variétés amusantes. Other plays written by anonymous authors dealt with the parachute invented and successfully tested by Jacques Garnerin, and also Garnerin's plan to even let a lady parachute down from a balloon. However, the police forbade the plan for moral reasons.

This French euphoria for aviation lasted for several decades. The French science theater tradition was continued in the middle of the 19th century by *Galvani*, the play by Antoine Andraud, and Louis Figuier's plays such as *Les Six Parties du Monde*, *Denis Papin*, *Kepler* or *La République des Abeilles* which were staged from the late 1870s on. Figuier strived with his "theatre des sciences" for didactical goals. In a period of nearly unquestionable confidence in human progress mediated by the sciences and technology, Figuier wanted to make the significance of science and also scientific results accessible to an audience of theatergoers.[5] Even if Figuier's ideas of the social function of science theater has become obsolete at least in part, French science theater is still alive today.

6.6 Science Theater Plays Authored by Scientists and Historians of Science

With the exception of *Infinities*, the plays we discussed so far are written by professional playwrights. Since science matters, or at least should matter, in a science play, playwrights like Michael Frayn and Penny Penniston did consult, for *Copenhagen* and *Now Then Again* respectively, physicists in order to avoid errors or misunderstandings concerning the physics in the play. The transition from plays stemming from the collaboration of a playwright with a scientist to plays written by scientists and by historians of science seemed quite natural, and it indeed happened in the last two decades. However, lacking sufficient familiarity with the realities in the theater world in most cases there are only very few within this group who reached a greater audience with their plays.

---

[5] For French science theater compare [Lachapelle, 2009].

This is certainly true for Carl Djerassi, a biochemist from Stanford University, who became famous and rich as one of the fathers of the contraceptive pill.

In contrast to Ronconi's and Peyret's conviction that today's science theater has to be stripped of everything biographical, in Djerassi's science plays science is closely interwoven with human emotions and actions. Djerassi is interested to introduce the theatergoer to the value system and the standards ruling the behavior of scientists in their laboratory work. In addition, he is interested to confront his audience with the social impact of scientific results and their applications. An example for this is his first science play *An Immaculate Misconception*, which was staged at the Edinburgh Fringe Festival in 1998 and had its premiere in a modified version at the New End Theatre in London in 1999.

The play deals with the ethical and social issues associated with contraceptive and assisted-reproductive methods. Dr. Melanie Laidlaw is a scientist in her late 30s. She tests the biomedical method known as intracytoplasmic sperm injection (ICSI) in a self-experiment by fertilizing one of her eggs with the stolen semen of her lover, Menachem. Her very ambitious partner at the lab, Frankenthaler, fertilizes without her knowledge one of her eggs with his own semen. Both eggs are later reinserted into her uterus. The result of the experiment is a baby called Adam. Adam is the source of all kinds of human complications, concerning motherhood and paternity.

In the science play *Oxygen* which Djerassi co-wrote with Roald Hoffmann, the Nobel Prize winner for chemistry in 1981, the two authors are concerned with the different claims of scientists and their reward system. In *Oxygen* three scientists of the 18th century who can claim to have discovered oxygen present their decisive experiments in a fictional encounter of 1777 Stockholm at the invitation of King Gustav III. They are the French chemist, Antoine Laurent Lavoisier (1743-1794), considered as the father of the chemical revolution; the English Unitarian minister and natural scientist, Joseph Priestley (1733-1786); and the Swedish apothecary, Carl Wilhelm Scheele (1733-1804). It was Lavoisier who understood the true nature of combustion, rusting, and animal respiration, and the central role of oxygen in each of these processes, formulated in the period 1770-1780. But it was Scheele and Priestley who discovered oxygen before Lavoisier, without understanding its function in the processes mentioned above. The play was set in 2001, the centenary of the Nobel Prize, when the Nobel Foundation decides to inaugurate a "Retro-Nobel" award for the true discoverer of oxygen. The questions addressed in this play are: What constitutes a discovery? Why is it so important to be first? Does it matter if you do not fully understand what you have found or if you do not let the world know about it?

The German translation of *Oxygen* appeared with a companion volume, to which I contributed an article on the priority dispute between Newton and Leibniz concerning the creation of infinitesimal calculus. Subsequently, this subject became the topic of other Djerassi's plays, *Calculus* and *Verrechnet*; the latter is different from *Calculus* and had its premiere on November 24, 2009.

Several of Djerassi's science plays, meanwhile eight, have been translated into other languages and enjoyed in part considerable success not to the least due to Djerassi's personality, especially his extrovert character. But even if Djerassi is an extraordinarily creative and successful scientist before he began to write his novels and science plays, it is impossible for obvious reasons to stage real science in action. So it is not astonishing that Djerassi began, like in *Calculus*, to take the stuff of his science plays from the history of science. But in the same way Djerassi had to learn to write like a playwright—more than twenty versions of *Oxygen* are witness to this—he had to learn how to find and to interpret the sources used by historians of science.

It is therefore only natural that historians of science took an interest in becoming playwrights for science plays. The most successful in this group was, or is, Marc Friedman who lived to see in 2003 the premiere of his one-act play *Remembering Miss Meitner* in three Swedish theaters. Friedman analyses Lise Meitner's contribution to the discovery of nuclear fission in the team of Otto Hahn who was awarded the 1944 Nobel Prize for Chemistry. To the team of Otto Hahn belonged also the chemist Fritz Strassmann. Despite the fact that both Strassmann and Meitner could claim a share in the discovery of nuclear fission it was only Otto Hahn who was honored with the Nobel Prize. In Friedman's play Straßmann in not even mentioned; Friedman is convinced that the Nobel Prize for chemistry in 1944 should have gone to both Hahn and Meitner. That she came away empty-handed Friedman interprets as a consequence of her fate that she, as an Austrian Jew, had to leave Germany in 1938. In addition Hahn appears in the play as the person who, by calling the Gestapo, forced Meitner to escape from Berlin to Copenhagen and later to Stockholm. In the play Meitner asks Hahn therefore in 1946 when he met her again in Stockholm where he received the Nobel Prize:

> Who, Hänschen, asked the Gestapo whether I would be allowed to continue my work? Who, Hähnschen, brought me to their attention?

In this way—unsupported by any historical evidence—Friedman alleged that Hahn have called the Gestapo to save his skin while pretending that he felt responsible for Meitner.

Even if a playwright is generally free to deviate from known facts and from well-founded interpretations, a historian of science who uses his subject as a playwright is not, if he claims like Friedman that everything in his play is based on historical fact. But still, Friedman was more applauded for *Remembering Miss Meitner* than his German colleague Falk Riess, who tried to present reconstructions of historical experiments in a drama setting. Like in Friedman's case, Riess's play *The Name of Fame* was never staged in Germany, with or without professional actors. Instead it was staged with the members of Riess' research team as actors, for a single time in 1999 for an exhibition in Antwerpen, Beligium and again in 2000 for the conference "Physics on Stage" organized by CERN in Geneva, Switzerland.

6.7 Conclusion

Since the beginning of the 17th century, the spectrum of science theater has ranged from refusal and fear of science to the conviction that an improvement of social and economical conditions can only be achieved by scientific progress. However, science theater in Germany could never claim any status except for a short period, during which it pointed mainly to the negative aspects of scientific activity.

It is important and correct to hint to the risks and perils of the applications of scientific results. But the science society of highly industrialized countries of the 21st century is neither willing nor able to step back to the technological standards of an earlier period such as the Pennsylvania Dutch who still live without electricity or automobiles. Generally speaking, today's science society cannot exist without the latest applications of scientific and technological research. Since science and technology permeate every aspect of our live they have become genuine parts of it. Therefore it is only natural that science has become an integral part of the stage where the human fate is illustrated with all its facets.

Science theater together with science movies and science television offers so much an opportunity to discuss critically and constructively the omnipresence of today's science in our world. Accordingly, ever since the late 1990s first of all in London and New York, the great interest in and going with it the success of science plays cannot be a short-lived fashion. In the USA, there exists a bunch of stipends and other subsidy tools for supporting young authors who aim to become playwrights for science plays or science movies; it promises bright prospect for science plays at least in the English speaking world.

References

Durrani, M. [2000] "Hawking slams 'stupid worthless' play," *Physics World*, August, 8.

Lachapelle, S. [2009] "Science on stage," *History of Science* **47**, 297-315.

Schneider, I. [2000a] "Ihr Auftritt, Mrs. Theorem. Warum Mathematik und Physik das amerikanische Theater erobern—und das deutsche nicht," *Frankfurter Allgemeine Zeitung,* August 3, 43.

Schneider, I. [2000b] "Ihr Auftritt, Mrs. Theorem," *Mitteilungen der DMV*, Heft **4**, 17-19.

Schneider, I. [2001] "Steven Hawking, Bühnenheld," *Welt am Sonntag*, January 7, 38.

Shepherd-Barr, K. [2006] *Science on Stage* (Princeton University Press, Princeton).

PART II

Arts in Action

7

Silence in Art

Cristina Leiria

In silence many things I have been finding and most of them are subjective, therefore resulting from my own experience. On those bases through the development of work started as an architect I have reached the sculpture world where I was given the opportunity to share an immensity of feelings related not only with the silence but also with light and peace. Through several stages of creativity we can be transported to new worlds.

7.1 Introduction

I, Cristina Rocha Leiria, architect specialized in Development Planning, has been working in sculpture and ceramics since 1993. My work is a form of art open to love and beauty, or at least it is my spirituality that enables me to conceive and produce this kind of art.

In this chapter, we can learn more about my experience as an artist through the presentation of my work, so we can realize the meaning of this message—*Silence in Art.*

Starting my professional life as an architect, I am now an architect with a sculptural approach and in recent years I have been devoting my work mostly to monument building and public art, trying to achieve my main purpose: the re-harmonization of the human spaces, both public and private.

7.2 Development of Sculpture Work

Through the development of sculpture work I was given the opportunity to share an immensity of feelings related with silence, light, and peace; therefore emerging an intrinsic need to share this entire message with the surrounding world:

> Learning to be grateful wherever good or bad experiences life presents, since they are all part of the universal equilibrium, I believe that everything in life comes for a reason. Through sculpture I could find the silence of my own essence.

Here a few works from Figs. 7.1-7.5 can illustrate what has been said concerning silence, light, and peace, and my intrinsic spirituality.

<u>Sail</u>

Just like a boat…
we are Body and Soul
Hull and Sail
sails tied to the mast, the wind blows
the boat adrift goes.
Sail in the wind
Faith, Trust and Hope
safely navigates the boat
and with honourable purposes
new and Vast Horizons are born.
Unfavourable wind blow
Helmsman of our own Life
in a spirit of adventure
sailing port-starboard
we reach the destination intended.
However …with a subtle movement of the Sail
in a gesture of Love, Truth and Courage,
respecting the surrounding world
we too, can change the Tack of
One's Life

Fig. 7.1. Left: *A Blue Waltzed Embrace* (2007), silver or silver plated, 31 x 8 x 41 cm. Right: *Sailing Boat* (2000), Bronze or Tinwork, 7 x 2 x 11 cm.

Family

In all sculptures entitled FAMILY, special treatment is given to movement, gesture, and its significance. In the approach to this subject there is the underlying idea of "universality." The observer may, from anywhere in the world and regardless of race, identify the theme and experience it in his own way.

This work (Fig. 7.2, left) may be seen from all sides, thus projecting the message to the outside and communing with its surroundings: relations, friends…universe. The enveloping and imaginary form is one of integration within an egg—the embryo potentially containing life and love. These are works of curving surfaces and soft contours devoid of angles, intending the nourishment of spiritual comfort to those who contemplate them. Love being the foundation of the Family, the sculpture is born of a heart, through whose lobes emerge the figures of father and mother. The feminine symbol represents succour, love, and intuition; the male symbol, uprightness, the protective instinct and reason. An imaginary spiral flows from the father and, passing through the mother thrusts itself into the cosmos through the child. With the conjoining of these figures in union, the intention is a representation of peace, simplicity, and harmony inherent to the true sense of Family.

Beyond

The Being discovers freedom, its true path,
In the hope of the Beyond…

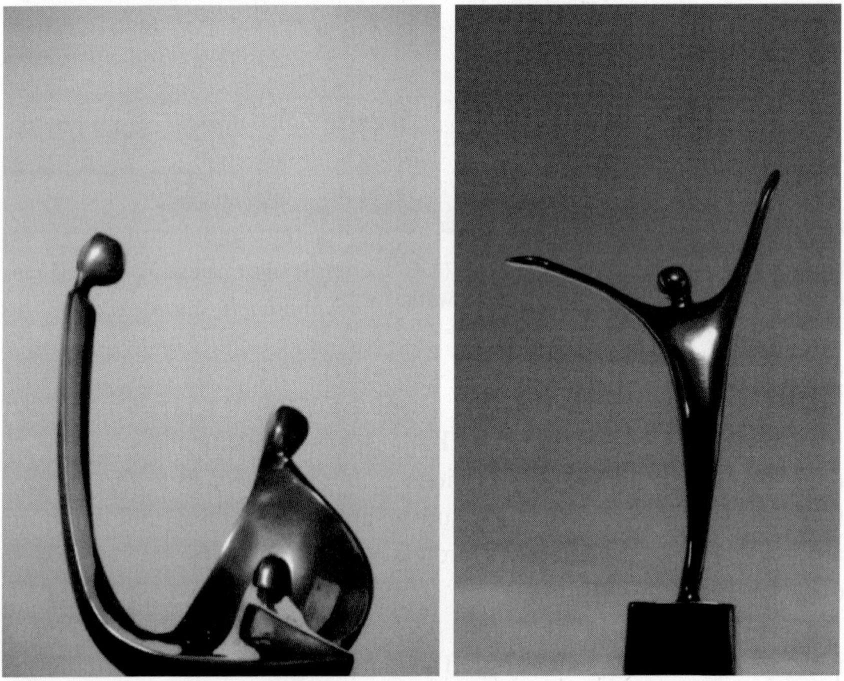

Fig. 7.2. Left: *Family* (1997), bronze with patina, 11 x 13 x 17 cm. Right: *Beyond* (2000), Tinwork, 3.5 x 3 x 20 cm.

Kun Iam

In the Orient, Kun Iam (Fig. 7.3) is considered a goddess of Love, Compassion, and Mercy. She is the caring Mother to the lonely, hope to would-be mothers, and protector of the pregnant and lovers; also attributed to her are miracles related to the sea and fishermen. In the east she is also appealed to in the quest for lost objects. Known as Avalokitesvara in India, Guann Yin in China, Kanon in Japan and Tchenrezi in Tibet, only in the south of China is she called Kun Iam. She is also venerated in the West as Divine Mother or Being of Light.

Fig. 7.3. *Guann Yin* (1999), 8.5 x 13 x 30 cm. [Manufactured by Crystal Atlantis.]

The Four Seasons

Four seasons, four moments of a life… four states of the soul… The four figures in Fig. 7.4 manifest the almost imperceptible change operating on all beings throughout the passage of time. In a harmonious and cadenced sequence, the energy contained in winter exalts in spring. In an ascending movement, arms girding the body rise to receive the first rays of sun. Hope is reborn and abundance emerges as the gift of summer. Arms open wide to give and receive, are now lowered to rest in contemplation and serenity of autumn. This pattern repeats itself eternally in nature and reproduces itself singularly in the life of every being. Thus, the joy of youth transforms itself into the energy of adulthood through the abandonment of a living maturity for the fusion of itself in the One when all is done. Each figure gently offers itself to the caress of sight and touch in a symphony of the senses and signs, reflecting the state of the soul which each one perceives, and which to some is given to express.

Fig. 7.4. *The Four Seasons* (1995), bronze, 3 x 1.5 x 11 cm. From left to right: Winter, spring, summer and autumn.

Saint Anthony's Tree

Saint Anthony retreats to his tree, where he builds a platform for prayer and meditation in the company of the Child Jesus (Fig. 7.5). Up above the dove illuminates, representing the Holy Spirit; two hearts, the symbol of love, outlines and includes the two figures. Several other works can be seen at: www.cristinarochaleiria.com.

Fig. 7.5. *Saint Anthony's Tree* (2004), brushed steel, 1 x 30 x 30 cm, 3 x 67 x 67 cm.

7.3 Creativity Stages

Here the different creativity stages of mine are presented in chronological order.

1960-1962: The Symbol

Born in Lisbon in 1946, at the age of two I was in Africa where I lived part of my childhood and youth, thus influencing my creativity. Between the age of 15 and 16 I created my first prototypes of two symbolic sculptures (Fig. 7.6).

Fig. 7.6. Left: *Presépio Holístico* (The Holistic Crib). Right: *Elan de Mãe* (Mother's Glow).

1963-1970: Contact with Reality and Emotivity

At the age of 16 I began my studies in Architecture at the Lisbon School of Arts (Escola Superior de Belas Artes). As a student, during the last years of the course, I worked on a project for a fire station (as a voluntary fireman) enabling an investigation on their working methods and needs.

At the end of my degree, I chose as thesis a project in social housing (Olivais Sul/Lisbon)—one already in use for 10 years. Following the new construction of this same project thus allowed me to understand the architect's intent and the way the plan for the existing space was used by the tenants. This experience originated a thesis which included a photographic and film presentation, and prospective users' recorded interviews on their degree of satisfaction related to the proposed facilities offered by the architect.

In 1970, already as an Architect, I cooperated with the Cabinet of Urban Planning and Architecture of Lisbon on an urban recovery project of the largest illicit district in Portugal, Brandoa, on the outskirts of Lisbon (Fig. 7.7). To fully apprehend the precariousness of the situation I cohabited with the residents, renting out a room in a ten storey building with no infrastructures: electricity, water, sewerage system or lifts.

Fig. 7.7. Brando, on the outskirts of Lisbon.

In this chaotic urban net, my presence as an "alien element" created, at the beginning, distrust and animosity but little by little it vanished. My intention as an architect was exclusively to help to improve the existing life conditions; the first objective was to ascertain, through district meetings, the most pressing needs. Thus began my collaboration in this urban project (which was presented by the studio I worked for, to the Lisbon Town Hall) and the urgent creation of the main urban infrastructures, including a proposal for the immediate construction of pre-fabricated pavilions to accommodate the Fire Station, Medical Services, Schools, etc.

1971-1973: Development Planning in London

After this Urban Planning project, I received a scholarship from Calouste Gulbenkian Foundation for a post-graduation study in the Development Planning Unit at London's University College. The purpose of this study was the application of urban planning to underdeveloped countries such as Mozambique. Later on I worked as an architect in United Kingdom (1973), Africa (1974-1977), Portugal (1977-1978), and Macau (1988-1990).

1974-1977: Africa Period

Mozambique, an underdeveloped country, needed a great support at the structural level once it had been devastated from a war period, leaving a ruined population. Concerned with this situation, I helped in the reconstruction of the country in collaboration with the Urban Cabinet of Lourenço Marques (Maputo) by creating nursery contact points, canteens, and schools for the native population. During this period I acquired a new experience concerning the common materials (such as wood, slum, clay, and straw) used in the performance of such constructions (Fig. 7.8).

Fig. 7.8. Construction in Mozambique.

The question of harmony related with urban planning had always been a focus in my architectural works. For example, those made in some neighbourhood countries of Mozambique, such as Rhodesia and South Africa, where I proposed architectural projects were based upon family balance and the surrounding spaces distribution, using some feng-shui theories.

1978- 1990: Portugal and Macau Period

In Portugal I developed an Integrated Plan for Almada region in collaboration with the painter, Eduardo Nery, enhancing features like

"Color and Texture" and including the surrounding landscape, therefore emerging a new concept of buildings. Later on other urban planning projects were developed within the Lisbon Urbanity Plan, giving rise to new architectural conceptions.

In 1988, Macao Government invited me to work as an urban planner architect. Therefore I felt my mission in Macau as being responsible for avoiding the destruction of beautiful gardens (mostly used by natives for Tai Chi practices) such as the S. Francisco Garden (Fig. 7.9) as well as the planning of other new urbanity projects.

During my stay at the Orient I profited from a Orient Foundation grant to study feng-shui and electromagnetic energy, giving rise to the organization of a new course for architects.

Fig. 7.9. S. Francisco Garden, Macau.

1990-1993: Portugal: The Transition Period

Other works had been performed during this period such as the construction of the Oeiras Pavilion. This event gave me the opportunity of future works at Oeiras Tourism Department, leading to the performance of touristic events.

1994-1996: Transition to Sculpture: Events Period

For health circumstances since 1992 a deep transition took place from architecture to sculpture, a new field for me; I created the *Angel of Light* (Fig. 7.10, left) in clay, as a therapy. Other sculptures took place following this experience; *Loving Birth* (Fig. 7.10, right) was later on manufactured by Crystal Atlantis, a Portuguese company.

Fig. 7.10. Left: *Angel of Light*. Right: *Loving Birth*.

A new philosophy of life based upon silence, flexibility, tenderness, love, and peace was developed through this experience since these are the main characteristics of my work, resulting in specific round, soft and elegant shapes. Most sculptures are made from clay and later on they were performed in several scales, in other materials such as crystal, bronze, silver, tin, and stone. Since 1995 some of these works of art, more than 200 originals and 27 sculptures have been edited by Atlantis/Vista Alegre, one of the most prestigious publishing companies. Here is the citation from Fernando António Baptista Pereira:

Her work grows among several issues such as: The Life, The Love, The Family, The Sea, and The Divinity; all of them linked together by an invisible line of universal coherence which is expressed by specific forms made up of the same language of love, balance, and peace. However, it is in 1999 with the emergence of Kun Iam Ecumenical Center Macau that the artistic experience created by Cristina reaches the highest level of congruence among the spiritual, social and architectural goals.

1997-2000: Macau's Sculpture: Kun Iam Ecumenical Center

Macau is both a Chinese and Portuguese city. The Portuguese proclaimed it "City of the Holy Name of God". The fishermen—the peninsula's original inhabitants—call it *Amacao* or, in Chinese, *Temple of the Goddess A Ma*. The followers of Buddha call it *Land of the Lotus*.

History can attest to the fact that Macau has always been a place where religious tolerance reigns. Hence, this territory is in itself an Ecumenical Center, opening to all philosophies and religions. The building, along with the statue of Kun Iam, symbolises what history has confirmed.

In 1997 I was commissioned to design the Kun Iam Ecumenical Center, located in the waters of the Pearl River[1] (Fig. 7.11). Luckily, my own philosophy always goes into the conception of my works; some of my sculptures have been created with a view to their possibly being turned into Public Art, to be placed in areas motivating humanization in urban living and, consequently, in city people's lives.

By Buddhists, Kun Iam is seen as a Bodhisattva (being of compassion and unconditional love, in the world, to free and save those who beseech her help), her life being spent in meditation on the sounds of the sea. According to feng-shui rules, being in water, Kun Iam should be facing the Land.

In my design, Kun Iam's face (Fig. 7.12) is done with an intention of internationality... no race evident. We can feel in it an expression of Asian, European, Indian and African features. The peaceful face is looking to a broad area where our eyes meet her tender sight. This area is

[1] www.macaumuseum.gov.mo/htmls/KunIam/story/story_eng.htm.

marked with a spiral rotating clockwise for a good energy to reach the place where people should stop and concentrate; this is the last symbol before entering the Lotus Flower Building. This spiral is part of a typical Portuguese stone pavement, on a bridge over the Pearl River.

Fig. 7.11. Kun Iam Ecumenical Center, Macau.

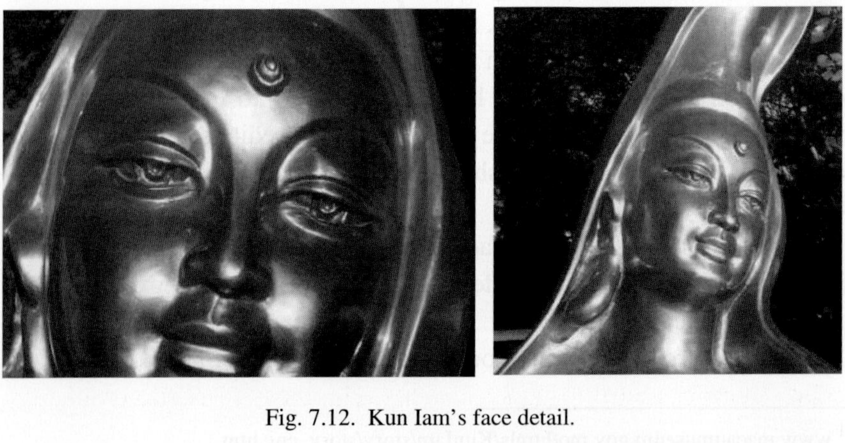

Fig. 7.12. Kun Iam's face detail.

The objectives proposed by the Kun Iam Ecumenical Center are in complete agreement with those of UNESCO—to forge peace in the human spirit. Not just because of its vocational aspect but also owing to its audacious artistry—crowned by a 20-meter high bronze statue of the goddess Kun Iam—this Center will certainly develop into a symbol of harmony and wisdom, especially for the young people of this world for whom it is destined. As with other monuments symbolizing different eras, the statue of Kun Iam will become part of the emblems identifying Macau and Southeast Asia. The citation from Helena Vaz da Silva describes it this way:

> While leaving Macau in 1999, Portugal wanted to leave a message of love, culture and tradition by offering a Portuguese interpretation to a Chinese divinity as well as a place created by a Portuguese artist to honour and respect Buddhism, Confucianism and Taoism principles, in support of Chinese civilization. This place promotes the meeting between the Orient and the Occident and is created by the Portuguese, the first Europeans to arrive China.

2001-2010: Monuments and Exhibitions

Beyond my performance of monuments execution, I had been invited during this decade to exhibit my work in several places. This enabled me to develop my artistic creativity by presenting new sculptures under the name of Public Art (Figs. 7.13-7.15)—from which we can see the humanization of urban life—and by creating special issues for specific commemorations such as those shown in Figs. 7.16.

Kun Iam

The essence of divine is one and lives inside of everyone. The Kun Iam (Fig. 7.13, left) sits on a lake at the House of Macau's garden in Lisbon, Portugal.

Saint Anthony and The Child

The child Jesus lies on the breviary, with a gesture of embrace to the saint; their eyes meet (Fig. 7.13, rgiht). The breviary is shaped like the prow of the boat which sank on the voyage of the young missionary on his way to North Africa. This was how "Saint Anthony of Lisbon" reached Italy and is subsequently also known as "Saint Anthony of Padua".

Fig. 7.13. Left: *Kun Iam and Lotus Flower* (2002), polished bronze, and metal painted in white, 2.7 m (House of Macau's Garden, Lisbon, Portugal). Right: *Saint Anthony and Baby Jesus* (2002), carbon steel, 2 m (Capuchos' Hospital, Lisbon, Portugal).

Mother and Child

To all mothers, a homage.
In attitude of reverence,
the Mother
raises her child to the Universe,
and with her gaze,
attends the heart
which beats within her.
The woman awakens the maternal being!
And between them
in joy,
the outline of a tear,

is born of womb and soul.
A singular gift,
the fruit of love.

Fig. 7.14. *Loving Birth* (2003), white-sea stone, 4.2 m (Municipal Park Marechal Carmona, Cascais, Portugal).

Sail

Fig. 7.15. Left: *Sail in the Wind* (2003), bronze statue, 2.7 x 1.3 x 12 m (Roundabout/Tavira, Portugal Centenary Celebration of International Sailing Federation (2007)/Special issues for that event). Right: *Sailing Boat*, bronze statue, 3 m high, over a marble wave (Cascais' Marina).

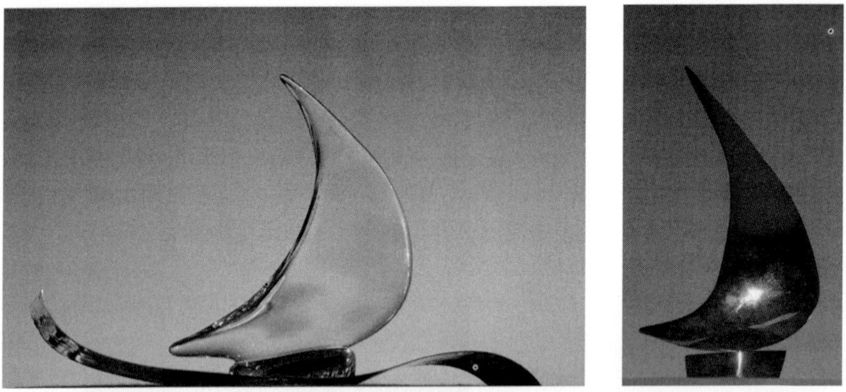

Fig. 7.16. Left: Trophy for the best world sailing team: Barbara Kendal, Ellen MacArthur, Eric Tabarly, Olin Stephens, Paul Elvstrom, and Robin Knox Johnson. Right: Sailing Federation Gift.

7.4 Conclusion

During the course of my life and from a very early stage, I was moved by situations of social differences and by the physical and psychological discomfort in existence in the inhabitants of the poorest zones of the urban net. Hence, my apprehension settle in and soon I begin to dedicate my architectural works to the comfort of housing, one of the essential foundations of social balance, defending salubrious spaces, harmonious forms, and integration in the involving environment.

My sensibility is also reflected in sculpture, as a gesture of freedom through simple and gracious forms in the search to transmit a message of well-being to those who contemplate them.

This is a participating method of helping human beings, while a "biological species", to grow better and go further. To conclude, as I said somewhere before,

> In silence many things can be found. Most of them are subjective, therefore resulting from our own experience.

8

Linsen's Art

Linsen H. Ngai

Linsen Hsia Ngai was a professional chemical physicist before becoming a self-taught watercolor painter about 25 years ago. Her selected artworks are presented here, in a chronological order.

8.1 My Background

I obtained a B.S. degree in Chemistry from the University of California, Berkeley, and my M.S. and Ph.D. degrees from University of Chicago majoring in Chemical Physics. After retiring from scientific research I became a watercolorist. I am a self-taught artist. I started out doing black and white paintings on Chinese paper and using Chinese ink. As I evolved as an artist, I began to use more color and less ink and eventually not using Chinese ink at all. I am still doing watercolor paintings on Chinese paper. My subject matter includes birds, animals, flowers, plants, landscapes, and figures and portraits. I want to show through my paintings the beauty of nature and to pass it on to future generations to enjoy.

8.2 Early Period

As a beginner in art my very first attempt was learning how to paint bamboo by using various shades of Chinese black ink and doing different strokes of bamboo leaves and stems (Fig. 8.1, left). Once I have accomplished the black and white bamboo paintings I expanded my paintings to other subject matters. All of my early paintings had a white background, but I realized in order to depict the diversity of nature I needed to incorporate color (Fig. 8.1, right).

Fig. 8.1. My early period. *Bamboo* (left) and *Searching for Food* (right).

8.3 Later Period

Bird Paintings

The next big adventure in my painting career was painting more elaborate feathered creatures with detailed backgrounds. My birds include ducks, geese, ibises, roosters, chicken, penguins, pelicans, and so on. I first started out painting only one bird (Fig. 8.2, left), then added two birds and eventually many birds (Fig. 8.2, right).

Love Birds

When I started painting my 2-bird art works I was actually thinking of them as love birds. Love birds to me also represent a loving couple. Throughout the years I have given my love bird paintings to marrying couples and for anniversaries as a celebration of love. For the love bird wedding gifts I usually wrote down the couples names, the wedding dates and the words 'wedding celebration' on the paintings that I presented to them. To me *love* is really the main bond of human society. As a result, I progressed into painting multiple birds (Fig. 8.3).

Fig. 8.2. My later period: Bird paintings. *Seeking for a Mate* (left) and *Two Cardinals* (right).

Inspiration from Others

It is a Chinese tradition to write poems, writings, comments, and verses on paintings. My father loved to write poems on my paintings. I remembered at one time I had discarded some paintings that I did not like. Then one day all of a sudden I discovered that my father had written poems on all of my discarded artworks. Because of his poems written on them I have treasured all these thrown away artworks.

I had also painted at one time 22 butterfly paintings. My father wrote poems on all of them and I did not even know. He then photographed them and made them into a small booklet that I entitled *Dreams of Butterflies* (Fig. 8.4).

Fig. 8.3. My later period: Love birds. *Black and White*.

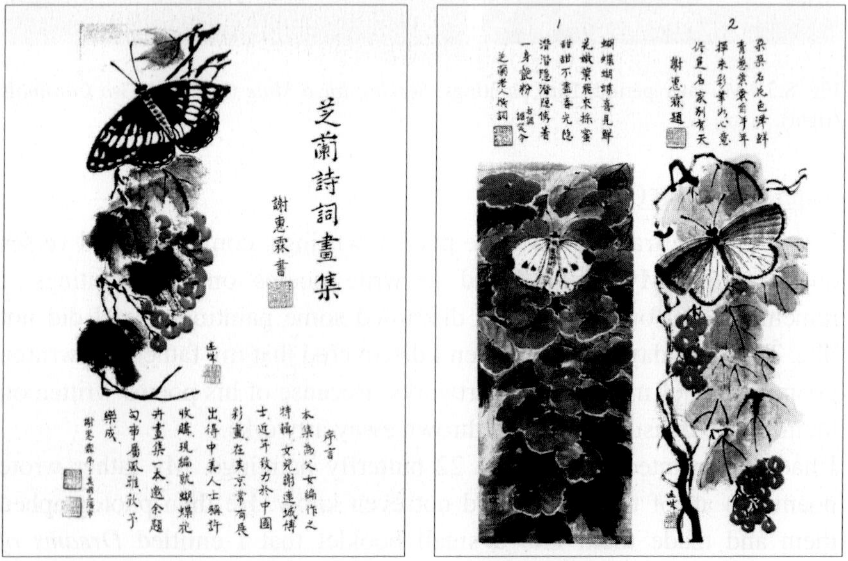

Fig. 8.4. Dreams of Butterflies. *Preface* (left) and *Butterflies* (right).

Because of this tradition of writing on paintings I also occasionally wrote something on my own paintings. In *High Perch* (Fig. 8.5, left), I wrote: "Yesterday I saw this bird perching high atop the rock."

About ten years ago the American Cancer Society had planted some experimental herbal plants known for fighting cancer at the Meadowlark Park in Virginia. All artists were encouraged to go there to paint the herbal plants. My two paintings were selected together with other artists' works to show throughout the United States for one whole year sponsored by the American Cancer Society. The purpose of this event is to show people a possible alternative way of treating cancer. The herbal plant in *Togetherness: I* (Fig. 8.5, right) is very interesting. It has red flowers, red stems, and even the veins of the leaves are red.

Fig. 8.5. *High Perch* (left) and *Togetherness: I* (right).

Fish to Frogs

At this point in time, my subject matter expanded from birds to fish and shrimp (Fig. 8.6), wild flowers and plants (Fig. 8.7, left), frogs and small creatures, wild animals, and still life (Fig. 8.7, right).

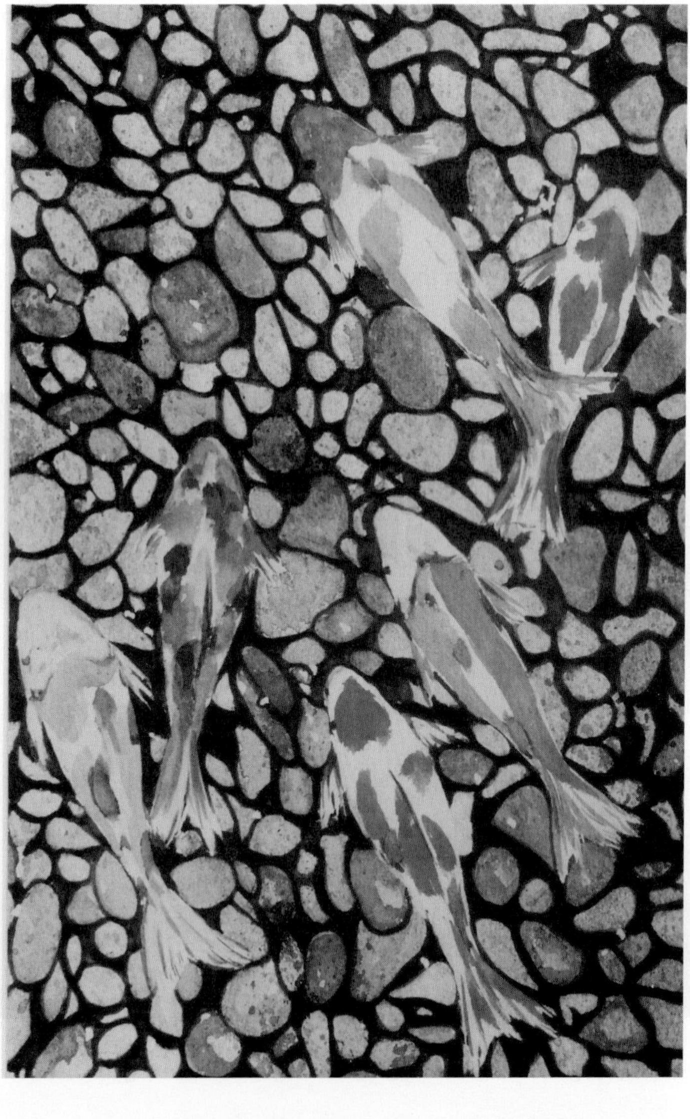

Fig. 8.6. My later period: Fish to frogs. *Colorful Mix*.

Fig. 8.7. *Whispering Whites* (left) and *Elk* (right).

Still Life

In *Autumn Tea* (Fig. 8.8), I wrote: "The tea is done; the chrysanthemums are blooming. Lovers of chrysanthemums, are you coming or not?"

Fig. 8.8. My later period: Still life. *Autumn Tea*.

From reading these few words on the painting one can appreciate the painting more: It is autumn; the host is preparing a tea party and has invited all the friends to come to enjoy autumn and to admire the chrysanthemum flowers grown in the garden.

Inspiration Abroad

The sources of my paintings of wildlife and plants are done out of my memories or from photos of nature. I picked a subject and composed it into my paintings. My other big series of paintings are landscapes. The subject matter is mainly of boats and lighthouses, the sceneries of the four seasons, and places that I visited with my husband in the United States and to foreign countries. I took a sketch book with me and did the scenery sketches with color pencils right on the spot. When I arrived home I did my paintings from my travel sketches. Here are some of my selected artworks (Figs. 8.9 and 8.10).

Fig. 8.9. My later period: Inspiration aboard. *The Blue Waltz* (left) and *Riverfront Houses, Germany* (right).

Fig. 8.10. *Cacti of Tucson, Arizona.*

Portraits

For some years I also worked on portraits and figures from life models (Fig. 8.11).

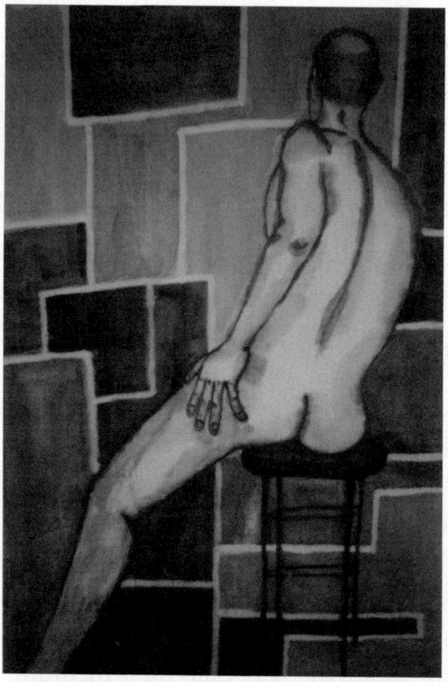

Fig. 8.11. My later period: Portraits. *Model on a Stool.*

Batik

The paintings presented above are all done on paper. Here, I would like to show some batik work I have done. Batik is produced on *fabric* with watercolor ink and wax. Batik is unique in the way beautiful lines are produced at random (Fig. 8.12).

Fig. 8.12. My later period: Batik. *Mother's Baby*.

8.4 Conclusion

Since I was educated as a scientist and worked as a researcher before I became an artist I have seen there are great links between (experimental) science and art. Experimental scientists work in a laboratory and do experiments while artists paint in their studios. Experiments take a lot of experimental skill to perfect and experimentalists connect research with scientific theories, while artists also spend lots of time to perfect their paintings, to accomplish their personal thoughts and ideas in their minds, and convey them in an artistic way onto their medium.

A painting can benefit from scientific knowledge. For realism, one has to know where the light of the painting is from, the shadow the light creates, and how the same color changes when light strikes an object. For example, a green color when strikes by a bright light may look like light green while the same green color in the shadow may look like a darker green and may have a bluish touch to it. If the same green color is surrounded by red or purple objects the green color would have some reddish tone to it.

For abstract art one has to know how to choose color so that the combinations of colors will make an impressive design on the whole painting. If one is working on a portrait the light will have many colorful alterations on the skin tone of the person. Realistic artist also has to know how to transfer a three-dimensional object onto a two-dimensional sheet of paper. The distance of an object is also very crucial in art. The objects in front are always larger than objects at a distance. All these are science, and yet all artists have to know about it and apply it to their artwork. Hence science and art are intermingling with each other all the time.

So I *do* think science and art are interrelated to each other, and they are part of the wonders of the world that we live in. We have to really admire and appreciate and enjoy them. I think this is part of life too.

I enjoy painting very much. Now I usually pick a subject and work on it for a while until I have decided to change to another subject matter and start working anew. I know there are still lots of rooms for improvement in my work. It is only by continuing to paint that I can improve my skills and better express myself in art.

9

From Curiosity to Creation: The Art of Holly Lane

Holly Lane

For the last several centuries specialization has caused science and art to diverge so as to seemly have little in common. However, a basis of commonality between art and science is the trait of curiosity, and both disciplines teach us to look attentively. The benefits science has bestowed upon the arts and humanity are acknowledged, and in turn the benefits the arts offer are cited. Reflections on the various definitions of art and speculations on the origin of art are offered from the perspective of an exhibiting artist. Some of the ways artists think about and approach, diverse kinds of art and the creative process will be discussed. Concluding, the artist's own work will be presented and prefaced by some of the concepts that shape the artwork.

9.1 Introduction

In the beginning, of human life, there was curiosity. In the beginning of each individual human life there is curiosity. Children are not embarrassed to ask questions, all kinds of questions. Even pre-articulate babies are curious; curious about balls of lint, noses, and the activity of the family dog. As individuals we are fortunate if, as we mature, curiosity is not squelched by fear, peer pressure, and inhibitions. If humankind was not curious we would probably not have science, art, philosophy, myths, poetry, books, electricity, medical cures, space exploration, zippers, and cheesecake, to name just a handful. Since there is so much to be curious about, to explore, individuals generally pick one category in college (or equivalent) that ideally, in alignment with their keenest interests, will fully utilize their innate talents and result in the flourishing of their minds. After accumulating knowledge, mastering skills and methods, gaining fluency in critical and creative thinking, and with curiosity still in tact, an individual has the potential to contribute to the advancement of human knowledge and well-being...and the well-being of the planet.

Scientists take one path and artists take another, but both share the trait of curiosity. The relation of science to art is that of allies of the good, with artists (and all humanity) benefiting from scientific discoveries and knowledge, and scientists hopefully benefiting from the ideas, insights, understandings, and experiences of art. In addition, I hope, art offers the solace of companionship, which in best instances is conveyable by art, the companionship of another human mind, that can stretch across the centuries through a work of art to a living person looking at it (or hearing it) today.

The next section (Section 9.2) is comprised of a freewheeling speculation on the origins of art; touching on recent archeological findings, an art historical re-evaluation of the Venus of Willendorf, animal tool use and the possibility of animal aesthetic awareness as clues to the beginnings of art.

The definitions of art posited by philosophers and thinkers through the ages are discussed from the perspective of a full time, exhibiting artist (Section 9.3). A few examples are given of a mindset of plurality and inclusiveness necessary in a world where various styles, from Primitivism to Postmodernism, can share the same exhibition space (Sections 9.4 and 9.5). In Section 9.6, the artist author/contributor's artwork is discussed and some examples shown.

9.2 Science and Art

Curiosity is a characteristic shared by scientists and artists. On a summer hike up a mountainside, one might just as well find an artist turning over rocks, as a scientist [1] [Sertich, 2010]. Added to curiosity, keen

[1] While hiking in Southern Utah, artist Joe Pachek discovered a dinosaur fossil and reported his find to the Utah Museum of Natural History. Paleontologist Mark Loewen and Anatomical Scientist Joseph Sertich, with the co-operation of the Bureau of Land Management, recovered the specimen; a new Basal Sauropodomorph Dinosaur.

observational skills and a predilection to think independently are other commonalities. Both science and art teaches us to be attentive, and see the world afresh. Curiosity impels scientists to discover the structure and processes of the knowable world through observation, reasoning and experimentation, employing the scientific method and using a wide variety of techniques and tools from dissection to linear accelerators. From this new basic knowledge, applied science creates new tools and processes. Curiosity in artists impels experimentation and the collection of disparate raw materials (visual details, symbolic import, psychological insights, color harmonies, shapes, etc.) to enhance, modify and orchestrate into a work of art, into something that never before existed.

For the sake of clarity, I will disclose that my starting assumption is that historically, science sought to discover, if not immutable laws of nature, at least principles that were reliable enough to make accurate predictions and be confirmable with repeated outcomes. Often the scientific inquiry sought to understand by reducing phenomenon to its essential components. New, productive concepts in science such as emergent properties, i.e., properties not evident from examining the individual parts, such as the fluidity of water is a property not evident when looking at H_2O molecules [Lam, 2008] have opened the door for the study of complex systems.

If one accepts the idea that science studies nature, and that humans are a part of nature, then it follows that the humanities are also valid topic study for science. As an artist I appreciate the recognition by science that art is a complex system. To illustrate, let us say with a very simple painting an artist is dealing with ten elements: six colors and four tonal values. The number of unique combinations are 45 variations, with the total number of permutations, according to a factorial calculation, being over three and half million, and this is not including the other fundamental components of a painting—line, texture, shape, volume, space, composition, subject matter, size, scale, etc.; the total possibilities are staggering.

In the last 400 years science has empirically proven itself, by offering not only reliable explanations that expand the knowledge of humanity, but also have made innumerable beneficial discoveries that have reduced human suffering. Science enjoys a well-earned credence, that, while not entirely supplanting the role of religion, philosophy and art (art in the form of story telling, myth making) as trustworthy explainers of the world, it has proven to be a reliable source of knowledge and method of inquiry. Daily I am grateful for the benefits science has conferred upon humanity: antibiotics, birth control, electricity, anesthesia, to name a few. Science has not only provided tools for artists, science ignites the imagination and deepens our understanding of the world showing us the richness of common everyday phenomenon as well as giving us a glimpse into the reaches of the vast universe.

9.3 Origins of Art

Art has been made by every culture in human history and robustly continues into modern times, in every corner of the world. Art today is often made with tools that are more sophisticated, but based on the same principles as early tools such as the chisel and mallet, airbrush, charcoal sticks, ground pigment in binder, and a paint "brush". It appears that art making is a universal of human existence; its practice and continuation implies that art is of great consequence to each generation of humankind no matter where or when individuals have lived.

In pondering about the origins of art, three obvious difficulties in pinpointing an accurate time and place, immediately come to mind.

1. Only artifacts or markings made from/of long lasting materials could have survived. Instead of rock or bone the first art could have been made out of leaves which would have quickly disintegrated. The survival of a fragile artifact would depend on the happenstance of being gently engulfed in amber, ash, tar or smoothly sandwiched in sedimentation, preserved in an intelligible condition and recognizable as anthropogenic. The first alteration to materials, made with aesthetic intent may have been earlier than the surviving Lower Paleolithic petroglyphs.
2. Speculation on origins hinges on how art is defined—the stretchy definition of art, will be further discussed later.

3. To know the mind or intent of a hominid 300,000 years ago is out of reach, but possessing brain capacity, dexterous ability, and tool use may be indicators that art making was achievable. Though the origin of art is currently an open question, some recent archeological discoveries, art historical hypothesis, and animal cognition research offer tantalizing clues.

9.3.1 *Archeological Evidence of Art*

While most art history texts cite the earliest art—recognizable to the modern eye—as dating between 40,000 to 10,000 years ago (e.g., the 17,000 year old cave murals in Lascaux, France), recent archeological research suggests that the first art making may have occurred much earlier. Small hemispheric holes laboriously pounded into quartz, granite and gneiss, termed "cupules" by archeologist Robert Bednarik [Hodgston, 2008] and numbering in the hundreds to thousands in certain locations, date between 700,000 and 300,000 BC. The social, cultural and aesthetic significance of these abstract carvings, made by *Homo erectus*, are not understood, but they required enormous physical effort due to the hardness of the rock, and appear not to have had a utilitarian use such as food grinding, climbing holds, storage, etc. The impetus to mark make, or make art continued to the Neanderthals. In the La Ferrassie Cave in France, 18 cupules were discovered on the underside of a limestone slab covering the grave of a Neanderthal child, believed to be 70,000 to 50,000 years old. While petroglyphs are the oldest surviving art, more ephemeral forms, such as portable carvings of bone, rock, wood, and cave murals dating back 70,000 years are further innovations in art making.

Archeology demonstrates that art making is not a static human enterprise, and that it is not the exclusive domain of one tribe, or one gender. After years of extensive comparative measurements, archeologist Dean Snow of Pennsylvania State University concluded that the proportions of the 28,000 year old stenciled handprints in the cave murals of Pech-Merle, France and Castillo, Spain were made by women artists [Snow, 2006; 2009]. The fact that the "spotted horses" paintings were created with the same technique as the surrounding handprints makes it highly likely that the paintings of animals were also created by women artists.

In addition, art historians are re-thinking other early art objects formerly assumed to be the work of male artists. LeRoy McDermott [1996] of the University of Central Missouri hypothesizes that the small, plump, female nude statuettes, created 22,000-21,000 BC (first discovered in 1908 near Willendorf, Austria by Josef Szombathy) and collectively named the Venus of Willendorf, are self-portraits.

9.3.2 *Tool Use and Aesthetic Awareness*

Though we do not have a written record of early hominid's thoughts articulating their artistic intentions, or how the objects were regarded in society, we can look for the presence of the basic preconditions such as tool use and aesthetic awareness, which are the prerequisites to art making. It is interesting to note that for the prospect of future evolutionary developments, both tool use and aesthetic awareness appears to be present in some animals.

Tool use may be a causative or concurrent evolutionary development along with art making. Neurologist Frank Wilson [1998] in his book *The Hand: How its Use Shapes the Brain, Language and Human Culture* proposes that use of hand tools spurred human brain development to its current large size.

It was once thought that humans were the only tool users but now the list of animal tool users is growing, including chimpanzees, gorillas, orangutans, crows, elephants, sea otters, octopi, macaques, bower birds, Egyptian vultures, striated heron, beavers, decorator crabs, archer fish, ants, wasps and dolphins. Michael Krutzen et al are studying a group of bottlenose dolphins in Shark Bay, Australia that use sponges, held in their snouts as they root around on the sea floor forging for food. In their findings [Krutzen et al, 2005] the researchers report that mitochondrial DNA analysis determined that all the "spongers" are the offspring of one "Sponger Eve" who taught her calves to use tools. Curiously, while both male and female calves spend the same amount of time with their mothers, sponging tool-use behavior appears to be matrilineal, transmitted from mother to daughter with only one male sponger recorded.

Tool use and the concurrent brain development suggest that animals might have the potential to become art makers. Admittedly, the other key precondition, aesthetic awareness is trickier to identify in other minds. Nevertheless, animals having an aesthetic awareness may be suggested by the peacock tail, as peahens' aesthetic preference for ever showier tails may have guided their selection of mates and therefore male evolution. Another hint of aesthetic appreciation in animals is suggested by wild animal behavior recorded by the pioneering psychologist Carl Gustav Jung. On a trip to Africa in 1925, Jung [1989] observed that:

> …big baboons would sit quietly, almost motionless on a high cliff and watch the sunrise every morning, whereas the rest of the day they ranged noisily, screeching and chattering. Like me they seemed to be waiting for the sunrise…

Though both Jung and the baboons sat quietly and watched the sunrise we can be confident that Jung was motivated by aesthetic longing and appreciation, because he could describe his thoughts to us. While we maybe inclined to explain the behavior of the baboons as biological, merely wishing to warm themselves in the rising sun light, an outside observer from another planet, who could not read Jung's account, could likewise explain Jung's behavior as merely an attempt to warm himself, absent any aesthetic motivation or awareness.

Biologist Sherman Bleakney [1970] outlined a promising shared evolutionary basis for aesthetic awareness in men and apes. Desmond Morris in 1966 observed that when presented with asymmetrical marked pieces of paper apes studiously balanced the symmetry by completing the pattern with appropriate brush strokes. Bleakney posited that since early humans collected natural objects possessing symmetrical or facial patterns, an artistic awareness of symmetry—possibly based on the survival necessity of an infant to identify its mother, blurrily keys onto the symmetry of faces—is a shared evolutionary development among primates and could indicate that animals indeed have the cognitive and perceptive prerequisites for aesthetic awareness.

While acknowledging that my conclusion is speculation, the new archeological finding about women cave painters and the discovery that a female dolphin transmits tool use to her offspring make it conceivable that female artists played a key role in the origin and perpetuation of art.

9.4 Definitions of Art

Definitions and theories of art have been offered by philosophers and thinkers down through human history from the Ancient Greeks to today; yet no one theory has been found to be entirely adequate [Titus, 1970].

The attempt to define art may be analogous to the story of the blind men and the elephant. Three blind men are touching a different part of the elephant. The blind man who is holding the tail announces that the elephant is thin and boney with stringy hair at the lower end. The second blind man who is touching the ear says, "No, the elephant is a big, thin, triangular flap." The third blind man holding the trunk counters, "No, the elephant is a long muscular tube that blows air." No one is wrong; they each unknowingly have a partial answer.

9.4.1 *Philosophical Definitions of Art*

A quick survey of philosophical definitions of art is presented here.

1. Plato's definition of art as imitation of nature is applicable to Realism, or representational art, but not necessarily applicable to other styles such as dematerialized Conceptual Art.
2. Art as expression of emotion is a workable definition for Expressionism, the Blues, and the love song or love sonnet but is not germane for Minimalism.
3. Art as pleasure, in common usage, "good" art, happy, entertaining art that is not intellectually challenging or gloomy, overlooks many masterpieces hanging in museums, paintings of war and martyrdom. Art as pleasure in the more specifically formal sense, as significant

form evoking aesthetic emotion [Bell, 1914], also excludes the modern developments of Art Povera, and much of Conceptual Art.

4. Intentionalism (or what was the intention of the maker—the artist's verbally stated intent determines all meaning) is an appeal to authority, leaves out the viewer, the artist's subconscious, and says nothing about the intrinsic qualities of the art object itself.
5. Immanuel Kant's idea of art as play, in contrasted to labor, suggested that the activity of art making is closer to play because it is an occupation agreeable in itself. While art making is deeply rewarding, it is a highly challenging activity, as it fully engages one's abilities and vivifies one's latent potentialities. However, in the idea of play there is a connotation of aimlessness, casualness, a diversion from important life issues that is alien to dedicated working artists. Much of art deals with important life issues; artists are thinking about and addressing in their work, environmental problems, socio-psychological dynamics, war, crime, justice, birth, death, suffering, and existential questions to name a few—art is not a diversion from, but a wholehearted involvement with important issues. To realistically test Kant's definition of art as play, simply announce that idea to an artist who has been sweating for months in her/his un-air-conditioned studio, working into the wee hours, employing back aching physical labor, expending a sustained mental effort in focused concentration, enduring considerable deprivation and gnawing uncertainty (since a work of art often has no guarantee that it will be purchased, and if it is purchased, be paid for in a timely fashion); tell that artist that s(he) has been playing and see what will happen.
6. Theory of art as empathy (catharsis included) is based on the sympathetic motor responses a person experiences when viewing for example, a Doric column whose vertical lines and solid girth gives the viewer a feeling of dignity and strength, diagonal lines convey speed and motion, horizontal lines evoke a calming effect. This definition, because it touches on formal qualities comes close to having a wide application, but it is partial as a definition for art, as one may have similar empathic response to a tree.

7. John Dewy believed that there is an aesthetic aspect to all normal experience, and in an attempt to de-spiritualize art he proposed that art should "idealize qualities found in common experience" [Dewy, 1934]. This definition is salient for Genre painting but is limiting as subject matter, as art should encompass everything, including the loftiest and rarest experiences, and states of being.
8. Art as communication (Leo Tolstoy being a proponent) and art as expression of intuition originated by Bennedetto Croce both point to powerful aspects of art: its ability to transmit thoughts, feeling, and even presence, compressing time, as we feel a presence—the reaching of a mind across the centuries—in the portraits of Rembrandt. The British philosopher G. E. Moore posited the idea that art is a thing in itself and needs no observers, appreciators to exist, or be valued; consequently, the art-as-communication criteria would not be relevant for Moore [Moore, 1903; Robinson, 2004].
9. A number of definitions of art separate it from craft on the basis of purpose—an object created for an utilitarian purpose is a craft, e.g., a spoon though exquisitely balanced and finely ornamented would be considered a craft object, not an art object as it has a function as an eating utensil, while a painting of an exquisite spoon would be art because the painting does not have a purpose, or function outside of itself. This distinction would make architecture a craft, to the consternation of many architects and a goodly number of art historians.

9.4.2 *Art as a Cluster Concept*

A cluster concept is an overarching concept that is composed of many characteristics, of which, any individual characteristic, or subset of characteristics is not necessary or sufficient [Gaut, 2000]. The cluster account of art can include all three blind men's definitions of the elephant, for it countenances a plurality of criteria. Rather than seeking one globally applicable principle that is solely necessary and sufficient for inclusion into the category of art, the cluster theory allows for plurality—an indeterminately large number of properties, and overlapping considerations. The cluster concept for art seems the most workable, as it is captious and among other things, has a theory of error, a normative dimension, explains why there is disagreement [Gaut, 2000], and it allows for the introduction of new criteria (which is essential as art is evolving, inventing new forms, and old forms are being re-defined as art). For example, a 19th century African mask made for ceremonial purposes maybe deemed as art in 21st century, due to its skill of execution and imaginative conception.

The four different aspects of the aesthetic experience that are emphasized to a greater or lesser degree in the various theories are:

1. qualities of the object
2. viewer's subjective response
3. the maker's intent
4. societal-institutional acceptance

They are all qualifying subset properties under the cluster theory, but no one aspect is sufficient, or necessary to qualify an object as art.

Some of the characteristics included in the cluster theory of art are: intellectual challenge, aesthetic pleasure, social relevance, layered meanings, emotional impact, symbolic weight, psychological insight, expression of intuition, communication of ideas, emotions, thoughts, extraordinary focus, significant form, skill, virtuosity, excellence, originality, inventive structure, iconographic references, narrative, coherence, uniqueness, metaphor, drama, eloquence, beauty, tragedy, conviction, ability to change or ennoble minds, continuity, ordered sound, balance, harmony, words, rhythm, movement, performance, creativeness, time, sequence, juxtaposition, proportion, mass, volume, shape, scale, color, texture, line, space, light, value (scale from black-grays-white), be human created, accepted by society, intentional, within a tradition, institutionally endorsed, power to move, and liminal-numinous-spiritual transport (Stendhal Syndrome[2]). A Rembrandt painting may have some

[2] Stendhal syndrome is a psychosomatic condition characterized by rapid heartbeat, dizziness, fainting, confusion and even hallucinations when an individual is exposed to art. Stendhal is the pen name of the French author, Marie Henri Beyle (1783-1842), who

of these characteristics but it is not necessary or sufficient that the Rembrandt painting have any particular one characteristic, or sets of characteristics.

9.5 On Thinking about Art

Since I came to the serious study of art from two years of philosophy I felt I had started with a gift; the position that every "isms", style and movement of art had something to teach about the nature of art. Being curious and therefore motivated to understand the different styles and movements (instead of judging as good or bad, right or wrong) proved to be sociologically and intellectually rewarding. If one accepts that there is no one right way to create, then as an evaluate criteria, one looks for artistic engagement, development of an idea or form, innovation, and the realization of an individual artistic vision. Either/or logic plays a reduced role in art making, while synthesis and bipolar logic[3] to name a couple, are more productive and fruitful modes of thinking. As a stretching exercise, when entering an art gallery I will look for a work of art that I do not immediately respond to, then I will study that work, until I understand it—sometimes this non-judgmental seeking phase lasts a year; when the "a-ha" experience occurs it is an exciting discovery. Art is endless; no matter how much one learns, there is still more to learn.

One of the benefits of being an artist is never being bored. Rather than seeing waiting-in-line at the post office as a boring task, an artist sees it an opportunity to study ears, or the backs of heads. Also, waiting in line at the post office provides a chance to study the folds in people's clothing, e.g., how pant cuffs buckle when they touch the shoe; a twenty minute wait flies quickly, thus employed. Instead of counting sheep as a calming technique to put anxieties out of mind before sleep, savoring and visualizing sculpting verbs like: dimple, speckle, pierce, flute, pucker, nudge, shrink, imbricate, twist, inflate, distort, stretch, juxtapose, extend,

---

described his own fainting experiences on a trip to Italy in 1817. It is said that Sigmund Freud fainted while viewing Michelangelo's marble sculpture of Moses.

[3] An example of either/or logic is when a person says "I do not want to participate because I will lose my individuality." Bipolar logic is operating when the person recognizes that the more he participates, the more individual he becomes.

gather, enhance, invert, orchestrate, multiply, scoop, mound, nudge, deepen, etc. for an artist is a centering, re-orientating exercise—a delicious work rehearsal for the next day's efforts.

Without exception, but still surprising, when I finish a work of art I get the oddest sense of pre-existence, that it pre-existed in another realm, and that stroke by stroke, chisel by chisel I, somehow, re-embodied it into this realm.

9.6 The Art of Holly Lane

After earning a M.F.A. I have been creating and exhibiting art in New York City for over 20 years. To illustrate the ideas behind my art, some of my artworks are presented in Fig. 9.1-9.8.

As an undergraduate I began to think about the nature and purpose of frames. Following Abstract Expressionism (when frames were completely discarded), a thin self-effacing frame made a re-appearance. Functionally, the thin frame was used to protect the painting, and it was required to be unobtrusive, to be invisible. The thin frames prompted me to think about frames as borders, as signifiers, delineating one realm from another. I began to see that the frame served as a signifier saying that all that was within the frame was art, and therefore should be viewed with heighten regard, and all that was outside the frame was not art. As I pondered these ideas, I ran across reproductions of medieval illuminated manuscripts in the University library. The illuminated manuscripts had elaborate borders with symbols, scroll work, birds, animals, figures and miniatures related to the text. Some of the borders were whimsical and commented on the text, even contradicting, or mocking the text. The idea that a frame could elaborate upon the enclosed text inspired re-conception of the frame as an informing context for my paintings. I wanted the self-effacing, "invisible" frame not only to become visible, but to become an integral part of the work of art. Since then I have been exploring the formal and conceptual inter-dynamics of frame and painting. With this approach a frame can be many things; the frame can be an environment, an extrapolation, an enhancement, a fanfare; it can expand upon the formal and conceptual aspects of the painting (Fig. 9.7); it can set up contrapuntal rhythms, extend movement, act as an architectural foil or translation (Fig. 9.5), be in conversation with the painting, housing the painting as a body houses and expresses the mind (Figs. 9.11, 9.4-9.7). With the frame as art, the border is removed, signifying that art is an embodiment of all that is, was, and may be in human experience.

The combination of *pictorial space*—the painting, and *sculptural space*—the carved frame, into one work of art provides the opportunity to simultaneously engage the viewer in two spatial realms is present. To inhabit the pictorial space in a two-dimensional painting the viewer must project her mind into the space, therefore I think of pictorial space as intellectual space. Sculptural space is physical space; the viewer approaches sculpture with an heighten physicality, proportioning her body to it, walking around it, sensing its mass and weight; therefore I see sculptural space as physical-body space. The two spatial approaches are simultaneously offered to create a complex aesthetic apprehension.

Some of my work includes small carved doors that open and close on hinges. This changeableness is usually a reference to time; behind the door is something in the future, something in the past, a cause, a consequence, or a hidden factor (Fig. 9.4).

Before majoring in fine art I studied philosophy. I switched my major to art because I longed to make ideas physical. I quickly discovered that art making did not involve an imposition of my ideas onto materials, but a collaboration with the physical qualities of materials; paint, graphite, canvas and wood had their own presence and potential. While I shaped the materials with ideas, the materials in turn shaped my ideas; this was and continues to be endlessly exciting and generative. In furtherance of my desire to make ideas physical (while continuing to make the painting/frame pieces) I have started to explore pure sculpture. I wish to see what sculpture's vocabulary of physical presence, mass, volume, form and space can articulate and embody (Figs. 9.2, 9.3 and 9.8).

Fig. 9.1. Holly Lane, *Summer Plans for Cognitive Augmentation* (2008), acrylic on wood, 44 ¼" x 37 ¾" x 11 ¾". [Photography by H. Lane.]

Fig. 9.2. Holly Lane, *The Well-Traveled Mind* (2006), gilded carved wood, 21 ¾" x 10 ½" x 8 ¾". [Photography by H. Lane.]

Fig. 9.3. Holly Lane, *After Dormancy* (2009), gilded carved wood, 66 ¾" x 9 ¼" x 9 ½". [Photography by H. Lane.]

Fig. 9.4. Holly Lane, *In Harmony with the Times, the Three Graces Take Public Transport* (2007), acrylic and carved wood, 48 ¼" x 37" x 8 ¾". [Photography by H. Lane.]

Fig. 9.5. Holly Lane, *The Long Awaited Day Finally Came* (2004), acrylic on wood, 24" x 23 ¼" x 9 ¼" (Private Collection). [Photography by H. Lane.]

Fig. 9.6. Holly Lane, *And So It Began* (2005), acrylic on wood, 66 ½" x 34 ¾" x 11 ¼" [Photography by Holly Lane.]

Fig. 9.7. Holly Lane, *The Root Seeker* (2005), acrylic on wood, 28 ¾" x 22 ¼" x 5 ¾" (Private Collection). [Photography by H. Lane.]

Fig. 9.8. Holly Lane, *Faculty Group Portrait* (2009), gilded carved basswood, 16" x 23" x 10 ¼". [Photography by H. Lane.]

Some of the themes in my work are: learning strategies (Fig. 9.1), growth after necessary dormancy (Fig. 9.3), the sociology of bus rides (Fig. 9.4), the reliance of human achievement upon the health of nature, proof of cognitive processes in animals, re-thinking myths by revealing what is hidden in a myth (Fig. 9.6), re-representing women (Figs. 9.1, 9.4, 9.6 and 9.7), unexpected correspondences, interspecies compassion, unobserved events, the topography of the mind (Fig. 9.2), the active and the contemplative life, planning for all contingencies, and a few honorific sculptures; to poets, teachers (Fig. 9.8), and for those who make gentle the life of this world.

9.7 Conclusion

Commonalities of scientists and artists are curiosity, keen observational skills, and a predilection for independent thinking. With science's new ability to study complex systems, the humanities and science may begin a new collaboration. Recent discoveries in archeology suggest that the origins of art may date back as far as 700,000 years, and included female artists. While different definitions of art emphasize particular aspects of art, the Cluster Theory stands out as most workable, as it does not seek one standard, but welcomes a plurality of criteria, any one characteristic or group of characteristic not being sufficient or necessary. Artists tend to think in a cumulative manner, using synthesis, incubation, and absorptive processes. I look forward to more curiosity and creation in the arts and sciences.

References

Bednarik, R. G. [2008] "The Origins of Symboling," *Signs* **2**, 82-113.

Bell, C. [1914] *Art* (Fredrick A. Stokes, New York).

Bleakney, J. S. [1970] "A possible evolutionary basis for aesthetic appreciation in men and apes," *Evolution* **24**(2), 477-479.

Chappell, J. & Kacelnik, A. [2002] "Tool selectivity in a non-primate: The New Caledonian crow (Corvus moneduloides)," *Anim. Cogn.* **5**, 71-78.

Dewey, J. [1934] *Art as Experience* (Putman, New York).

Gaut, B. [2000] "'Art'as a cluster concept," in *Theories of Art Today*, ed. Carroll, N. (University of Wisconsin Press, Madison).

Hodgston, D. [2008] *Pleistocene Palaeoart of the World*, ed. Bednarik, R. G. (British Archeological Report, London).

Jung, C. G. [1989] *Memories, Dreams, Reflections*, ed. Jaffé, A. (Vintage, New York).

Krutzen, M., Mann, J., Heithaus, M. R., Connor, R. C., Bejder, L. & Shervin, W. B. [2005] "Cultural transmission of tool use in bottlenose dolphins," *PNAS* **102**, 8939-8943.

Lam, L. [2008] "Science Matters: A unified perspective," in *Science Matters: Humanities as Complex Systems*, eds. Burguete, M. & Lam, L. (World Scientific, Singapore).

McDermott, L. [1996] "Self-representation in Upper Paleolithic female figurines," *Current Anthropology* **37**(2), 227-275.

Moore, G. E. [1903 & 1988] *Principia Ethica* (Promethus Books, New York).

Titus, H. H. [1970] *Living Issues in Philosophy*, 5th Ed. (Van Nostrand, New York).

Robinson, D. N. [2004] "Aesthetics: Beauty without observers," in *The Great Ideas of Philosophy*, 2nd Ed., Parts I-V, an audio book by Robinson, D. N. (The Teaching Company, Chantilly, VA) pp. 292-293.

Sertich JJW, Loewen MA [2010] "A New Basal Sauropodomorph Dinosaur from the Lower Jurassic Navajo Sandstone of Southern Utah," *PLoS ONE* **5**(3) doi:10.1371/journal.pone.0009789.

Snow, D. [2006] "Sexual dimorphism in Upper Paleolithic hand stencils," *Antiquity* **80**, 390-404.

Snow, D. [2009] "Pictures: Prehistoric European cave artists were female," *National Geographic* (online pictures), June 16.

Wilson, F. R. [1998] *The Hand: How Its Use Shapes the Brain, Language and Human Culture* (Vintage Books, New York).

10

Making Movies and Making Physics

Hark Tsui and Lui Lam

The characteristics and experiences of making movies and making physics are discussed, respectively, by a movie director/producer and a physicist. Similarities and differences between the making of movies and the making of physics are presented. Discussions on the nature of movies and physics, on creativity and innovation as well as on the joy of making movies and making physics are provided.

10.1 Introduction

Making movies is a creative process that involves several operational stages: (1) conception of the project, (2) lining up the funding, (3) finding coworkers, (4) shooting the movie, (5) post-shooting work, and (6) marketing and distributing the movie.

Making physics means creating new physics—doing physics, or physics research, at its best. It is also a creative process and, like making movies, involves the same six stages in its operation, from the beginning to finish. The exception is that when doing pure theory or simple experiments, stages (2) and (3) may be absent.

Hark Tsui has been directing and producing movies since 1979 [Ho & Ho, 2002; Morton, 2001] and Lui Lam published his first physics paper in 1968 [McMillan et al, 1968]. Presently, Tsui has directed/produced over 70 movies and Lam has published over 170 research papers and 12 books. The two knew each other since 1975 when both were doing community work in the New York Chinatown, in lower Manhattan.

In the following, their background and their views on the nature of movies and physics, respectively, are given in Sections 11.2 and 11.3. Their experience in making movies and making physics—picking a project and executing the project—are presented in Sections 11.4 and 11.5. Some musings on creativity and innovation (Section 11.6) and the joy of working on movies and physics (Section 11.7) are also provided. Section 11.8 concludes the chapter with discussion.

10.2 Our Background

A person's background influences his personal and professional choices and working style, which shows up in the final product this person produces, whether it is his movie or physics. The backgrounds of the authors are therefore given here first.

Tsui: I think it is very difficult to trace back one's background relating to one's creativity, because most of the time ideas come from factors related to DNA instead of environmental influences. That is why some people can draw when they are not trained as a painter, and some can sing when they are not trained in music academies. It is very abstract and difficult to link up someone's creativity to his/her growing process.

My understanding of what initiated my creativity happened in my childhood in Vietnam, starting with my sketch of a drawing. In fact, something happened before that, when I was younger. A school classmate drew a chalk drawing on a small blackboard for fun, which somehow was related to my later inclination of becoming a movie director. That drawing is about a movie that I saw before in a cinema, *Godzilla Attacking Tokyo*. In that experience I was quite frustrated because I could not draw as good as he did. In fact, I did not draw at all. But watching him draw was great fun. In fact, my classmate's drawing activity was accompanied by singing and talking. It was very much like what one experienced in a cinema and watched animation done by those classmates.

Very soon after that, I picked up a pencil and tried to draw, and it was the copy of a car show's pamphlet. It was great fun. This first trial of my creativity energized my curiosity to find out what I can draw. Then to surpass my classmates, I started to draw movie shoots. I went through the period when a lot of kids played with the flashlight shining through a plastic with drawings, with narratives—a basic kind of movie.

Later, I switched my interest to drawing buildings, which made my family believe that I was going to be an architect. It was very encouraging that my family elders asked me to draw different buildings and awarded me with petty money. Those drawings included people with activities inside buildings, a very basic video game like Sim City. The whole idea of using projector, drawing and narrative to tell a story was put aside when I entered the preteen period. It was later in my teen years, I, with a few friends, rented (daily when needed) a regular 8 mm camera, with my own pocket money and everybody chipped in.

At that point I moved to Hong Kong. For the first three years school life was unlike what I had before in Vietnam; in Hong Kong, it was total book reading and many exams, only a little bit of drawing. With this trend continuing, my interest shifted to *physics* and science in general, wanting to be a scientist, being influenced by a society that respects scientists. Those years were spent in reading a lot of science books, gearing to be a scientist. Graduating from high school, I had to decide what to do in college and in the future—quite a funny idea for a 17-year-old schoolboy; he had to make such an important choice without enough knowledge. It was a big decision to make because going to college took a lot of money, and I had no money; I had to think hard.

At the end I decided to switch to humanities and be happy. An interesting question arose: how to be happy while working as an adult. For two years I hanged around, searching for a decision. Suddenly one day on a bus, a friend asked me, "What will make you happy?" My answer: see movies, rather than working. My friend, "Then why don't you make movies?" Very inspiring conversation. After that day, I asked myself what to do if I wanted to be a movie director. With the limited knowledge I had at that age, I asked for advice. Most people went to study film or media production with ready financial support or some basic knowledge of the field. I had none. (My family did not see any promise in filmmaking and so did not really support my ambition.).

Bring the little bit of money I earned in high school, I went to the United States with enough money for staying one semester in college only; I was constantly frustrated by the idea that I might be forced to return to Hong Kong soon. But the experience of being in a foreign country and living independently was very exciting. Every stage happening in life seemed to worth my effort. New perspective of what the world looked like opened up my horizon of thinking. The possibility of having to return home again did not stop me from trying my best to stay in the US. I tried very hard to do what I thought was needed to stay in college to learn what to do as a filmmaker.

The first Christmas, I went to New York looking for a job. Then, after staying in college for three years, in 1975 I went to New York again working in a film-processing lab. I was very much into New York and involved myself in community life—where people lived in Chinatown, midtown, with different cultures. There were so many things to do and to see; everything became lively and hopeful. It was very satisfactory seeing myself as a useful person.

An essential factor for me going to New York was to do a paper that was needed in my courses. The plan was to do a documentary film in New York, about the community and the society, a film about antiwar activities. I was looking for a documentary film person I had in mind but could not find him. I ended up working in a documentary company myself. Later, my involvement in the Chinatown community helped me to develop further. For example, while in New York I went to the Chinatown to see a movie from China, *Tianshan Red Flower*; the movie was in English while the audience was Chinese. Instant translation was needed, but no one could translate it well into Chinese. So I volunteered, and that happened again and again. The Chinatown turned out to be very complicated, with many groups of different interests coexisting. I looked for different channels to help people—nonprofit organization helping the locals to find jobs, Medicare clinics and media production groups.

In 1977 I returned to Hong Kong in order to understand my race and my mother country at a close distance, since I could only get a remote feeling of these things in New York. Hong Kong was then the closest place one could get near the Mainland. I wanted to wait in Hong Kong for China to open its door so that I could enter Mainland, to explore my experience as a Chinese and my heredity. While in Hong Kong I tried to do documentary films, but was instead asked by my employer to make

TV films. Fortunately for me, the experience helped. I directed my first feature film *The Butterfly Murders* in 1979 and became a movie director ever since.

In the Mainland in 1991, I made my first film *Once Upon a Time in China*; it was a very dramatic experience. The movie systems in Hong Kong and in Mainland were not the same, not even similar to each other in mentality; it was very disturbing to me. But then those valuable experiences would ready me to work in China on a long term basis later. Seeing those experiences as the preparation for future challenges, I continued to think of how to become a filmmaker in China.

In the mid-1990s, there were a lot of invitations from Hollywood for moviemakers in Hong Kong. I became one of the chosen directors to go; I was back to the US again. Then I was offered the chance to stay in the US for good, an exciting chance indeed. But I opted for Hong Kong, more precisely, China, since Hong Kong is part of China. I wanted to see, as a filmmaker, whatever would happen in the 1997 transition when Hong Kong was reverted back to China, to open up my own horizon as a filmmaker in my own country.

My plan of making movies in China continued. However, there were a few proposed projects that did not get through the censors in China. So my dream of being a filmmaker in China was temporarily hindered. In 2002, I was involved in a company that wanted to produce a movie called *Seven Swords*. And I launched myself again, on the track of shooting a movie with permissible content. Since then, I had been in China, thinking of all possible projects for this particular market.

However, SARS (severe acute respiratory syndrome) suddenly broke out in 2003, in Hong Kong and elsewhere. My attention was turned back to Hong Kong after much of my personal relationships there had been neglected by me. I was thinking that so much more could be done for Hong Kong, by injecting creativity in that society to create more interesting culture and present them to the rest of the world. I started planning a community project called Project Hong Kong, to extend the local culture and increase the appeal of Hong Kong to the outside world. The project did not go well because of insufficient financial and social support; it was a nonprofit project, supported neither by the government nor private corporations. For two years, there was no result and I became losing interest in this project.

In 2005, *Seven Swords* was finished. I did other movies in China, like *All about Women* and *Missing* (both 2008)—the latter was shot in Taiwan, Hong Kong and Japan—and *Detective Dee and the Mystery of the Phantom Flame* (2010).

Lam: I was born in Guangdong Province and grew up in Hong Kong, where I received my education from grade one on and graduated from the University of Hong Kong with a B.Sc. degree, spanning from1949 to 1965. I then went aboard to Vancouver, Canada, in 1965 and then to New York City in 1966. I received my M.Sc. degree from the University of British Columbia and Ph.D. from Columbia University (1973), both in physics.

It was during my graduate student years in the late 1960s that the anti-Vietnam-War movement erupted and the students took over the buildings at Columbia University. From end of 1970 to 1971, I actively participated in the *Baodiao Movement* [Lam, 1971]; [1] and then lived in New York's Chinatown to serve the community for about two years. We started the "Chinatown Food Coop" and helped the local patriotic newspaper *China Daily News* in its publication.

When at Bell Laboratories in Murray Hill, New Jersey, as a graduate student from Columbia, my mentor was Philip Platzman. [2] I did my

---

[1] In 1970, after large quantities of potential oil deposits near the group of tiny islands called Diaoyudao (called Tiaoyutai in Taiwan) were announced by foreign oil companies, both Japan and China reiterated their ownership of Diaoyudao. To help keep Diaoyudao under China, overseas Chinese students in USA started the "Protect Tiaoyutai" (*Baodiao*) movement at the end of 1970 [The Seventies Monthly, 1971]. The first large gathering was held in the basement of the College of Education building at Columbia in Dec., 1970. Subsequently, many oversea Chinese, students or otherwise, worldwide were mobilized. See, e.g.: http://archives.lib.nthu.edu.tw/exhibition/diaoyun/ (April 27, 2009).

[2] Phil Platzman did his Ph.D. thesis with Murray Gell-Mann in particle physics, and polaron work with Richard Feynman at Caltech before he worked at Bell Labs, the place he spent his whole career life. Both Gell-Mann and Feynman (1918-1988) are Nobel laureates. In 1997, Platzman and Peter Eisenberger shared the APS Arthur H. Compton Award (http://www.aps.anl.gov/About/Committees/APS_Users_Organization/Compton/index.htm, June 20, 2010).

postdoc under Melvin Lax (1922-2002)[3] at City College, City University of New York (1972-1975) before moving to Antwerpen, Belgium and then Saarbrücken, West Germany, spending about one year at each place. Starting January 1978, after the Cultural Revolution, I worked at the Institute of Physics, Chinese Academy of Sciences, in Beijing [Lam, 2010]. I left China at end of 1983 due to family reasons [Li, 1983]. Subsequently, I worked at City University of New York (1984-1987) and at San Jose State University since 1987.

Essentially, my physics research ranged from nuclear physics in my Vancouver years (1965-1966) to condensed matter theory at New York and Europe (1969-1978), and then from liquid crystals research in China and New York (1978-1987) to nonlinear physics and complex systems in San Jose (since 1987). I started publishing on the humanities in 2002 [Lam, 2002; Lam, et al, 2010].

I do not recall that I was interested in how the physical world works when I was young. I got interested in physics research after I published a few papers on my own as a graduate student at Columbia. However, I do recall that in my high school years, probably grade 11, I wrote a short essay in class under the title, "Those Who Don't Think Ahead Will Have Recent Worries," which is a well-known Chinese idiom chosen by the Chinese literature teacher as the essay title. Instead of putting out examples that illustrate the correctness of this statement like my classmates did, I gave the mechanism that guarantees this phenomenon. The mechanism goes like this: The world is very complicated (I did not know the term "complex system" yet), and so troublesome things will keep on appearing. If one do not think about them in advance and plan on how to handle them, they would become recent worries when they do happen. I wrote it short because the mechanism is simple. And instead of getting a high mark like 90% or more that I expected, I got a pretty low

---

[3] Melvin Lax, one of the founders of quantum optics, received his Ph.D. from MIT (1947) and worked at Bell Labs (1955-1972) before assuming the Distinguished Professorship at City College, CUNY. He was awarded the Willis E. Lamb Medal for Laser Physics in 1999 (photonics.usask.ca/interestingtopics/files/Laser%20Invention/Melvin_Lax.pdf, June 20, 2010), and elected to the National Academy of Sciences in 1983 (www.aip.org/ history/acap/biographies/ bio.jsp?laxm).

grade for my perfect argument. Looking back, this is my first research work on *Science Matters*.[4]

10.3 What Are Movies? What Is Physics?

Tsui: Movies present a phenomenal mentality shared by the masses. I always believe that to understand movies, one has to understand the history of movies. From movies' history, for example, one can understand why some movies are made with certain social factors. Also, to study (or to create) a classic movie that is enjoyed and appreciated by people for a long time, one must has a registration of the feeling or emotion of that certain period in time the movie depicts. To understand or revisit a certain era, one can go back to see old movies. Our obligation as a movie maker is to be sincere about how we feel about current events happening in that era, to give a phenomenal mentality shared by all.

Movie is a mirror from which the audience can see themselves.

Lam: "Physics is what physicists do." This is the definition of physics held to by *Physics Today*, the official monthly published by the American Physical Society [Lubkin, 1998, p. 24]. Indeed, this is the only definition that makes sense since the domain of physics is ever changing. Mature areas like classical Newtonian mechanics shifted from the physics to the engineering department, and new areas such as *econophysics* [Mantegna & Stanley, 2000] and *histophysics* [Lam, 2002] came in.

Today, physics is not just about nonliving or simple systems; it is also about complex systems, including all those from the social sciences and humanities. As advocated by the Nobel laureate Arthur Schawlow: "The task of physics is not only to understand the hydrogen atom, but to understand the world."[5]

---

[4] Science Matters is the new discipline that treats all human-related matters as part of science [Lam, 2008; 2011].

[5] Quoted in *Physics Today*, Oct. 1994, p. 53.

10.4 Picking a Project

Tsui: Every filmmaker always goes through history of his own growth from the childhood days, from time to time. When watching old movies, the filmmaker may come up with a new perspective and new ideas to replace the old ones—integrating it with new moralities, viewing it from a new angle, and using a new story-telling method. This is because only a member of the new generation can grasp the mentality of that new generation. New direction of movies always comes up and keeps on happening. Consequently, "remaking" a movie is not necessarily, and usually not, a simple remake of the old movie; the remake is not the same movie.

To choose a topic as a contemporary movie maker, one should make connection with the environment and the world, and look for passion and emotion that can touch you and other people. This is the basic rule in picking a project in movie making.

As a naturally born artist, the movie maker always gets nutrition from what urges him on to express himself in how he sees the world. In most of the time, the ability to express himself comes from the blood of the filmmaker. For instance, some directors may not learn story telling from academic channels; yet, they still can tell a very touching story. That is why sometimes it is DNA that makes a person what he is.

Lam: To pick a project in physics research involves two steps: (1) coming up with an idea; (2) deciding whether to go ahead with the idea. There are three ways that a good idea may come to mind, like the case that a photographer may capture a good picture.

1. She could sit still in front of her house, with the camera in hand, and capture whatever that is interesting and happening within her eyesight.
2. She could run after a rushing crowd, join them and see what is happening out there; she may get a good picture of something exciting.
3. She goes places and sometimes wanders around; she may capture good pictures from time to time.

Needless to say, method 1 is not recommended because of the slim chance of success. Method 2 corresponds to chasing a hot topic in physics research like high-T_c superconductivity when it was first discovered in 1978. The competition will be keen and one should go into it only if one thinks she has a good idea in solving the problem or enough resources that could beat out all the competitors. Method 3 is most productive and is usually preferred.

However, with method 3, there are still two problems to overcome. First, like the photographer with the camera, the research physicist should be equipped—with research skills ready or capable of picking up necessary skills quickly. Second, how does she know where to go? The solution to these two problems comes from the same source: accumulating information in your brain on a *long-term* basis—since the research topic and needed tools are not known beforehand. To do that, it involves (1) reading the monthly magazine *Physics Today*, (2) browsing all the physics journals in the library, and (3) attending weekly departmental seminars.

Reading for research is very different from reading in a physics course. Read only the abstract as well as the first and last sections of an article is enough; go back to read in detail when you need to use the material there someday with your research topic in mind. And in a seminar, paying attention to the first and last five minutes of the one-hour talk is essential, because the speaker, the expert in that topic, is summarizing for you in the first five minutes what have been done on that topic and why the research is worth doing, and in the last five minutes the new finings and open problems. Most significant works are achieved by borrowing concepts or tools from one research field to the other; completely new things are invented on very rare occasions. That is why one has to read books and journals and attend seminars which do not fall neatly into the research topic one is doing. Do not let your mentoring professor tell you otherwise.

When you have a choice in picking problems, aim high. Pick the one that will have an impact, but you still have a fair chance of solving it. To that end, you have to *guess* the level of difficulty of the solution, what tools are needed, and what your own level of expertise is. In short, you have to know yourself and your available resources pretty well; make friends and connections all the time because they are part of your resources. Of course, to play it safe before you have a tenure job, you may want to work on a familiar problem that guarantees publications while you are tackling a difficult but exciting one. The trick is always have an exciting problem under working, even if that means you have to hide it from your mentor who is using you as a cheap labor.

Wondering around from time to time, i.e., trying something new and unexplored by others, is equally important because important discoveries often show up unexpectedly. This is best described by Alexander Graham Bell (1847-1922), the inventor of the telephone: "Leave the beaten track occasionally and dive into the woods. You will be certain to find something that you have never seen before" (Fig. 10.1).

Fig. 10.1. Bust of Alexander Graham Bell at the entrance of Bell Laboratories, Murray Hill, New Jersey. Bell's quote (see text) is inscribed below the bust. [Photography taken by Lui Lam, July 12, 1994.]

10.5 Executing the Project

Tsui: The production of a movie involves more than one person; it is a team work. As such the director is the commander of the team who has to lead the way in thinking. The intention of the director has to be made clear to everybody so the group would follow the intended direction and come out with a work that has a solid vision.

When working with a group, the director's demanding control is necessary to avoid diversion of different ideas. The director has to be firm in his stand, even to confront people who do not agree with him, like a commander in an army. This is due partly to time and financial constraints, but mostly to artistic control.

It is quite often that the director participates in writing the script and draws his own storyboards (Fig. 10.2), and involves himself heavily in post production works. All of these, naturally, are for the director to express himself as an artist and for the artistic control of the final product, the movie. Lastly, one should not forget that the making of a movie involves not just the production component but also the financing component; a movie is a "commercial" product, in the positive sense of the word. That is why the director and actors have to participate in the promotion of a movie before or after it is finished, to help the sale of the movie.

Lam: A researcher has to be *completely* honest, trustworthy and dependable; never work with anyone who does not meet these criteria. The reason is very simple: There are many steps in a piece of research that it is sometimes quite impossible for others to check the details such as an extremely long and complicated calculation, the writing of a computer program, and the taking and analysis of data. Moreover, it may take a year or more to accumulate data and do the analysis, and if this person lost the records of them it will be a waste of valuable time for the whole team. In research, priority is of paramount importance; there is no second runner.

Good physics research involves two parts: the major and important part comes from guessing and imagination, or what one calls *intuition*; the other part is *logical*, coming from induction or deduction. Intuition cannot be taught as simply as logical deductions; intuition is what differentiates good physicists from less physicists. When one faces a new phenomenon, the first step is to guess what happened. Modeling (Fig. 10.3) or equation writing comes later; proofs of any kind come even later and are often done after the correct answer is first guessed.

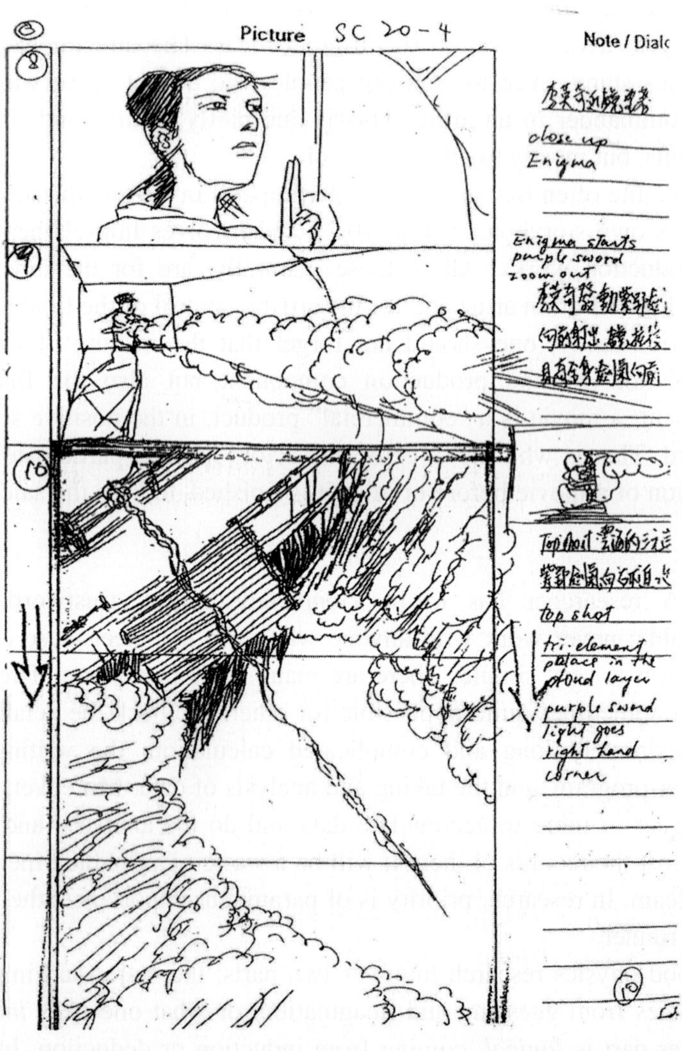

Fig. 10.2. Storyboards for *The Legend of Zu*, 2001 (drawn by Hark Tsui).

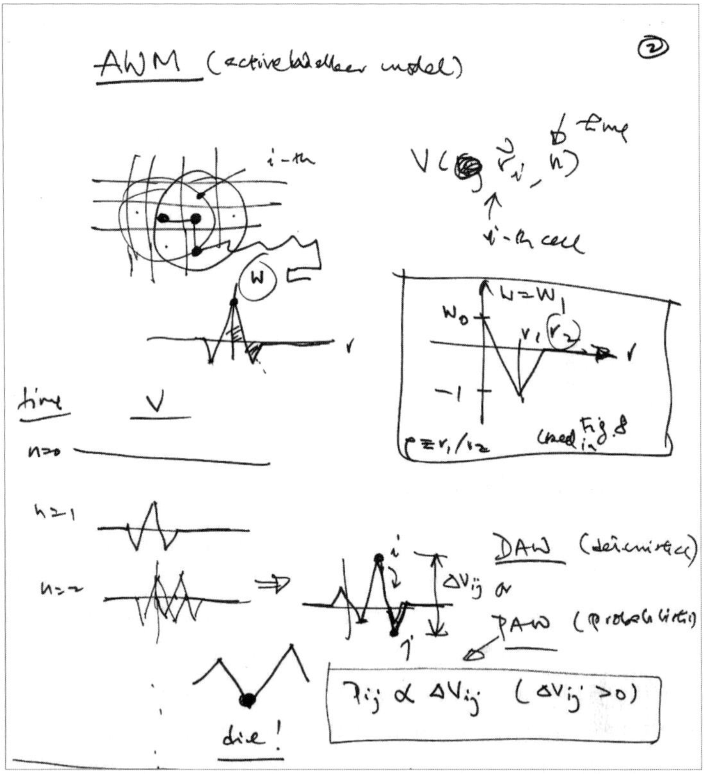

Fig. 10.3. Constructing an Active Walk Model (drawn by Lui Lam, March 21, 1992). The AW model was initially designed to reproduce the filamentary patterns observed experimentally, but turned out to be a general paradigm covering many complex (human and nonhuman) systems [Lam, 2005; 2006].

Publishing a paper is the byproduct of research. The basic aim of research is to *understand nature*, not to publish papers—a point often misunderstood or ignored by many practicing physicists, which, unfortunately, might be beyond their control. Counting papers by administrators indicates the lack of qualified referees in the evaluation process. The publishing of fraudulent papers actually misleads the research direction and thus hinders the development of a research field. Anyone knowingly committing fraud and any institution tolerating fraud among its members have no business in research whatsoever. And, in fact, reporting fraud in one's own field is considered part of the duty of the researcher, irrespective of the researcher's motivation.

Reporting one's research results in seminars or conferences is part of the game, which could happen before or after the paper is published. The motivation is about advertising one's own work, but more importantly, it is to ascertain and improve one's research by facing challenges from one's peers. In physics research, while the picking and execution of the project, and even the formulation of one's results and the writing of the paper—those parts involving human action—are human dependent and could be the domain of sociology of science, it does not follow that the contents of the end scientific results, like Newton's laws of motion or Einstein's relativity theories, are also human dependent. In fact, they are not; they belong to the human-independent knowledge [Lam, 2008].

10.6 Creativity and Innovation

Tsui: One never knows a priorily whether he has creativity or is capable of innovations. The director has to come up with the result of his work—the movie, to test with the audience by finding out the direction of their response so that the director will know what he did is indeed what he wants. For example, one might want to do a very touching story, but the resulting movie may end up with a totally different effect. The reason could be that the materials of his work clash with his method of story telling, the two working in opposite directions. A creative person always has to explore different possibilities which might bring him to unknown realms.

The unknown realm is always a mysterious space of self exploration, which sometimes could be very risky and uncertain. For instance, good directors do make films that end up as crab; all filmmakers do make films that the audience does not accept. Those films in the course of self exploration may not be mature enough to establish communication between the director and the audience. Sometimes it may take more than one step to get to the destination the director wants it to be. And it is the filmmaker who has to fight all the wars to get to the point that he can successfully come up with, with good creativity. A director always has to focus on his audience of his choice.

Innovation is not just about techniques. It is a vision you want to pass to the audience. To achieve that, the director has to choose all the "weapons" accessible to him.

Lam: Creativity means the ability to create something novel. It involves three components: dare to think, free to speak out one's mind, and the tolerance of failures. There are material bases behind all these. Every child dares to think but such an innate ability might be suppressed by the society, at home and in school, when the child grows up. To be a good physicist, one should be able to look into Einstein's eye and tell him he is wrong, if that is what one thinks. But this is possible only if there is a feeling of "equality" among the two persons engaged in an academic discussion, *and* any displeasure of the senior person will not jeopardize one's immediate position and future career. Furthermore, failures occur all the time during a creative process; that is why innovations are so hard to come by. The tolerance of failures is thus very important and the society must provide multiple ways for a failed person to come back, e.g., freedom to switch majors in a university, the ease of changing schools, and the existence of different types of colleges (such as the two-year community college and the four-year university) that is linked to each other. In short, we are talking about the freedom to think and speak, and the flexibility and mobility in the education system and the employment circles.

Providing pleasant and conducive working environment is equally important in encouraging creativity and innovation. Sitting the graduate students in small computer booths in a bare room is not the best arrangement. For comparison, at California's Pixar Animation Studios where award-winning movies are produced year after year, the working place is designed like a playground almost like the Disneyland.

10.7 The Joy of Making Movies and Making Physics

Tsui: Every time one goes for the process of self exploration, one finds more perspectives of oneself. It is more than an excitement; it is the means to see that there is always more about yourself that you do not know about. There are always unknowns out there waiting to be discovered.

Lam: The joy of doing physics comes mostly from practical rewards: flexible working hours, reasonable income, job security and freedom to work on any topic of your choice (if you are a tenured professor), friends and colleagues around the world, and the chance to visit beautiful places (where conferences are held). Moreover, if you work on pure physics, you may discover one or two fundamental laws of nature with profound implications that might change the world—a job satisfaction hardly matched in other professions. The dream of winning a Noble prize and the belief that you are discovering "truth" about the universe and so on help, but are not required. Lastly, if you are curious about many things in the world, including humans, and think hard enough, physics research does provide a unique training that helps you to see things with a unified perspective.

10.8 Conclusion

We both had received training in the United States but had worked for many years in environments with less or limited resources. Yet, we have always tried to compete and, whenever possible, outperform the best in the rest of the world. In this regard, we hope that those in similar situations will find something in this chapter useful.

Human affairs are stochastic systems that involve probability considerations. There are always exceptions to the rules. What works for us may not work for others, vice versa. However, since the objects and subjects involved in movies and physics are both part of the natural system, with constraints such as physical principles and human nature ever present, it follows that there could be more similarities than differences between movies and physics, or, at least, in movie making and physics making.

References

Ho, Sam & Ho, Wai-Leng [2002] *The Swordsman and His* Jian Hu: *Tsui Hark and Hong Kong Film* (Hong Kong Film archive, 2002).

Lam, L. [1971] "Hong Kong, Taiwan, Diaoyutai," *New York Hong-Kong-Students Monthly* **20** [reprinted in *Truth Behind the Diaoyutai Incident*, ed. The Seventies Monthly (The Seventies Monthly, Hong Kong, 1971) pp. 103-105].

Lam, L. [2002] "Histophysics: A new discipline," *Mod. Phys. Lett. B* **16**, 1163-1176.

Lam, L. [2005] "Active walks: The first twelve years (Part I)," *Int. J. Bifurcation and Chaos* **15**, 2317-2348.

Lam, L. [2006] "Active walks: The first twelve years (Part II)," *Int. J. Bifurcation and Chaos* **16**, 239-268.

Lam, L. [2008] "Science Matters: A unified perspective," in *Science Matters: Humanities as Complex Systems*, eds. Burguete, M. & Lam, L. (World Scientific, Singapore) pp. 1-38.

Lam, L. [2010] "The first 'non-government' visiting-scholar delegation in the United States of America from People's Republic of China, 1979-1981," *Science & Culture Review* **7**(2), 84-94.

Lam, L. [2011] "Arts: A Science Matter," in *Arts: A Science Matter*, eds. Burguete, M. & Lam, L. (World Scientific, Singapore).

Lam, L., Bellavia, D. C., Han, X.-P., Liu, C.-H. A., Shu, C.-Q., Wen, Z., Zhou, T. & Zhu, J. [2010] "Bilinear Effect in complex systems," *Europhys. Lett.* **91**, 68004.

Li, Yuan-Yi [2003] "The unbroken China complex: The life story of a Chinese scientist," *ScienceTimes*, Aug. 8.

Lubkin, G. B. [1998], "A personal look back at Physics Today, *Phys. Today*, May, 24-29.

Mantegna, R. N. & Stanley, H. G. [2000] *An Introduction to Econophysics* (Cambridge University Press, New York).

McMillan, C. D., Bhargava, P., Lam, L. & Vogt, E. W. [1968] "A calculation of the production of pions by 450 MeV protons on nuclei," *Can. J. Phys.* **46**, 1141-1144.

Morton, L. [2001] *The Cinema of Tsui Hark* (McFarland & Co., Jefferson, NC).

The Seventies Monthly (ed.) [1971] *Truth Behind the Diaoyutai Incident* (The Seventies Monthly, Hong Kong).

11

A Journey along the Borderland: A Critical Approach to Artificial Intelligence-Based Art and Literary Practices

Jichen Zhu and D. Fox Harrell

Digital media open up opportunities for new integrations of art and science. However, the close contact between the multiple cultures in either tradition also unveils fundamental value differences that impose considerable difficulties in performing interdisciplinary work. In this chapter, we identify a new form of this cultural divide in the context of computer art and digital media practices. Next, we identify a growing number of practices that engage the capacity of the computer to abstractly represent data and to process it algorithmically in order to serve expressive, critical, and generalizable purposes. In particular, we explore artificial intelligence-based art and literary practices that actively negotiate the hidden assumptions and push the disciplinary boundaries of both art and science. Finally, we present our AI-based interactive narrative work *Memory, Reverie Machine*, which engages literature, cognitive science, and AI. It illustrates our perspective and strategy of combining art and science practices synergistically, as part of a growing community for which the exploring of the borderland between art and science can transform not only particular technologies or how they are perceived, but also end goals and values.

11.1 Introduction

Half a century ago, British scientist and novelist C. P. Snow [1964] delivered a now notorious talk on the increasing gap between the two cultures of the sciences and the humanities [Lam, 2008]. Reflecting on his personal and social experiences as a scientist and novelist, Snow pointed out the lack of communication fueled the deep cultural divide between the two major intellectual campaigns. According to Snow, the resulting miscon-ceptions and distrust was counterproductive to solving real-world problems.

To those who were concerned with Snow's intellectual divide, the advent of digital media brought new opportunities to bridge the gap. Indeed, in the last decades, the humanities/arts and science/engineering communities have developed and shared numerous common approaches to the computer. In areas concerning screen-based content, practitioners constructed metaphors of screens as "pages," widely used in both Web art and scientific data archives, or the screen as a "canvas," in animated digital painting and information visualization in computing research. Some see computer as a tool, for example, in the creation of synthesized music as well as the design of a new chip layout. Others may conceive it as something for humans to relate to [Turkle, 2004], in the form of either cajoling video games characters or medical robots that treat ailments and attend to patients' emotional needs. Finally, thinkers and practitioners from both cultures may place their focus on computer algorithms and/or knowledge structures, as the cases in software art and artificial intelligence research.

Despite what may seem like a new Renaissance age, however, the close contact between the arts and sciences[1] does not automatically eliminate the fundamental differences in their values and methodology. For instance, to offer an over-generalized summary, goals of the arts often revolve around issues such as aesthetics, expression, and/or social critique, whereas the sciences still typically value utility, empirical understanding of the world, and/or productively generalizable insights.

On the contrary, we argue that uncritical crossovers between the two cultures may polarize their differences and deepen the divide. In Snow's time, intellectuals ignored each other. "Oh, those are mathematicians! We never talk to them" [Snow, 1964]. If the gap fifty years ago was mainly due to the lack of interest and basic literacy in the other culture, it takes its shape as the battle between conflicting core values and beliefs. Unable to reconcile these differences, many early digital art-combine-

---

[1] We will use the umbrella term "sciences" to refer to the related communities of sciences and engineering, and "arts" for the arts, humanities, and design.

with-science experiments fell short of their interdisciplinary claim by merely scratching the surface of the other culture. They simply regarded the other cultures as merely a tool for accomplishing work in the domains of their native cultures. This cultural imperialism has intended consequences. For instance, uncritically using simulation technologies without taking into account its deep root in military trainings may undermine ethical claims of an art piece [Penny, 2004]. Similarly, trying to artificially constrain digital artwork to recognizable art world conventions, e.g., hanging framed computer screens as new computer "paintings" and ignoring century-worth of discourse on paintings and recent theory of the unique characteristics of software art as distinct from that tradition may be a dangerously traditionalist approach.

In this chapter, we examine the mutation of Snow's two cultures in the context of computer-based art and cultural practice and articulate our strategies of combining art and science practices synergistically as peer practices. In particular, we explore artificial intelligence (AI)-based art and literary practices that actively negotiate the hidden assumptions and push the disciplinary boundaries of both art and science. In the rest of the chapter, we will first identify representations of the two cultures in contemporary digital media practice. We will then discuss existing approaches for which the intersections between art and science can transform not only particular technologies or how they are perceived, but also end goals and values. In dialogue with notions such as Michael Mateas's "expressive AI" [2001] and Noah Wardrip-Fruin's "Expressive Processing" [2009], we engage the capacity of the computer to abstractly represent data and to process it algorithmically in order to serve expressive, critical, and generalizable purposes. Next, we will present our AI-based interactive narrative project *Memory, Reverie Machine*, which engages literature, cognitive science, and AI. Finally, we will discuss our strategies of combining art and science practices synergistically.

11.2 Integrating the Two Cultures

Today, the intersection and exchange between the two cultures take place frequently in the realm of digital media. An increasing number of practitioners from each side attempted to reach out to the other side of the divide. However, the distance between the two cultures does not seem to be disappearing as many people predicted. In the rest of this section, we will paint a broad picture of the contemporary manifestations of both campaigns in the context of computer-based art and digital media, and why we found certain attitude towards interdisciplinary work problematic. We will also offer what we consider as more fruitful approaches for engaging and bridging the two sides of the divide.

11.2.1 *"Duchamp-Land" and "Turing-Land"*

Parallel to Snow's two cultures, digital artist and theorist Lev Manovich [1996] depicted a similar divide in computer arts between what he called *Duchamp-land* and *Turing-land*. The former refers to the community of galleries, major museums, prestigious art journals, whereas the latter describes a more technology-focused art world, exemplified by major venues such as ISEA (Inter-Society for the Electronic Arts), Ars Electronica, SIGGRAPH art shows. In spite of their overlapping interest in computer arts, the two lands pay homage to very different intellectual traditions. Duchamp-land, named after the master of modernist avant-garde art, is a continuation of the traditional fine art practices. It orients itself towards "content" and multiple cultural codes, often with an ironic, self-referential, and destructive attitude.[2] Turing-land, on the other hand, gravitates towards the latest, state-of-the-art technology, and is frequently quite self-reflexive, engaging and exploiting the nature of the technologies themselves.

As computer art evolved, venues became more accommodating in letting in different approaches. But the cultural divide between the arts and sciences persisted. Like Manovich, Simon Penny [2007] observed a similar polarity in works that address both computing and the arts across disciplines:

[2] Some destructive art practices that Manovich refers to include the Self-destructive Machines by Tinguely and the first exhibition of Nam June Paik "where he screwed technology—ripping open television sets or changing TV signals by affixing magnets to the monitors."

> A significant difference between computer science research and media arts practice lies in the ontological status of the artifact… for an artwork, the effectiveness of the immediate sensorial effect of the artifact is the primary criterion for success. It is engaging, it is communicative, it is taken to be coherent, or it is a failure. The criterion for success is performative. Most if not all effort is focused on the persuasiveness of the experience. Backstage may be a mess, a kluge. In computer science the situation is reversed. If the physical presentation is a little rough around the edge, or even missing entire pieces, this can be overlooked with a little handwaving, because the artifact functions as a "proof of concept" which points to the real work, which is inherently abstract and theoretical.

These different approaches and perspectives, Penny argued, are deeply rooted in the core ideologies of the two cultures. Science's insistence upon "alphanumeric abstraction," logical rationality, and desire for generalizability are fundamentally in conflict with the affective power of artwork, which is based on specificity and complexity.

When Snow made his observation, the two cultures would not and could not talk to each other because practitioners well-versed in one rarely had basic level of literacy in the other. Fifty-one years later, computational literacy is much more widespread. Many exciting explorations and collaborations are taking place across the cultural divide, and new interdisciplinary areas are emerging. However, we also need to be aware that the conflict between the two cultural ideologies has grown arguably more intense with this close contact.

Our major concern is that some of these collaborations and interdisciplinary inquiries are motivated by an implicit "cultural imperialism," instead of healthy, informed exchanges. As we pointed out earlier, some practitioners from either side saw the other culture as merely a foreign land that their native culture is set to conquer. Under this mindset, some computer scientists see arts as an application domain to which their algorithmic framework can be applied to; likewise, some artists regard computers as merely a tool to achieve their unaltered visions, without understanding or questioning the worldviews that these "foreign" elements embody. In regard to uncritically adopting ideologies from the traditional science world to computing, and particularly to art practice, Penny warned us of the danger of the "unquestioned axiomatic acceptance of the concept of generality as being a virtue in computational practice … especially when that axiomatic assumption is unquestioningly applied in realms where it may not be relevant." Similarly, attempts to reproduce art as a scientific experiment without an in-depth understanding of its values and discourses are equally simple-minded or even costly.

11.2.2 *The Borderland of Critical Computing*

Though the phrase has been used previously, Harrell's particular notion of critical computing is in the context of producing what he calls "phantasmal media"—most simply described as those computational works that engage human culture, ideology, and conceptualization through algorithmic and data-structural means [Harrell, 2010]. In short, phantasmal media use computing for subjective, cultural, and critical aims. Regarding critical computing in particular, he states that it:

> … refers to the potential for using algorithmic processing and data structuring as expressive bases for expressing commentary about, and making impactful change upon, the world of human experience. The critical computing concept helps technologists to move beyond development of utilitarian and productivity-oriented applications [Harnell, 2010].

Though here critical awareness is directed both externally toward the world and internally toward technology itself, a major inspiration for this perspective comes from self-critical approaches to computing and information sciences. This is perhaps best exemplified by the work of Philip Agre [1997a; 1997b] that addresses the confrontation between the two intellectual traditions in the contemporary research area of artificial intelligence (AI). As a branch of computer science dedicated to the formal study and production of human-level intelligence through computer algorithmic operations, AI also directly engages many long-standing concerns in the humanities tradition, such as the nature of intelligence and intentionality, and sparked debates in boarder contexts. In the political context of a rising Cold War [Edwards, 1996], its ambitious goal and discursive power not only attracted many computer scientists, engineers, psychologists and military funding agencies, but also quickly engaged philosophers, critical theory scholars, practicing artists, and popular culture producers, who were concerned with the nature of intelligence and its implications of human identity. Some multi-disciplinary debates of AI seemed like battles between parallel universes, unable to establish a shared intellectual space and time; but others led to fruitful exchanges and reflections [Suchman, 1987; Weizenbaum, 1976; Winograd & Flores, 1986]. Among them, Agre's *critical technical practice* [1997b] has far-reaching impact in other technological fields beyond AI —for example Phoebe Sengers's research applies his ideas to highly original approaches to Human-Computer Interaction (HCI) designs using computing within everyday experiences and environments [Sengers et al, 2004].

In his attempt to reform the field, Agre turned to the critiques of AI from various humanities fields—phenomenology, literary theory, and anthropology. At first, from the perspective of a more traditional AI practitioner, these texts seemed impenetrable, and hostile. As described by Agre [1997b]:

> All critical analysis may seem like accusations of conscious malfeasance. Even sociological descriptions that seem perfectly neutral to their authors can seem like personal insults to their subjects if they presuppose forms of social order that exist below the level of conscious strategy and choice.

After considerable cultural shocks, self-reflection and adaptation, the first breakthrough came when:

> [I]t finally occurred to me to stop translating these strange disciplinary languages into technical schemata, and instead simply to learn them on their own terms. This was very difficult because my technical training had instilled in me two polar-opposite orientations to language—as a precisely formalized and as impossibly vague—and a single clear mission for all discursive

> work—transforming vagueness into precision through formalization… I understood intellectually that the language was "precise" in a wholly different sense from the precision of technical language.

The gradual internalization of these foreign works provided new vocabularies, methods, and perspectives to scrutinize the hidden assumptions and worldview related to science and engineering, often taken for granted by its practitioners. At the core of his argument, Agre challenged the long-standing premise that science and engineering are objective and value-neutral, and pointed out different "ideologies and social structures [embedded in AI research, that are] … reproduced through a myriad of unconscious mechanisms such as linguistic forms and bodily habits." The goal of a critical technical practice is thus to "cultivate awareness of the assumptions that lie implicit in inherited technical practices" [Agre, 1997a].

Agre's work demonstrates the importance of constructive collaborations between the two cultures. It inspired a new community of critical technical and technical critical practices, including, among others, Sengers's AI agent design informed by culture studies [Sengers, 1998], Michael Mateas' expressive AI [Mateas, 2001], Ian Bogost's procedural rhetoric [Bogost, 2006], Noah Wardrip-Fruin's expressive processing [Wardrip-Fruin, 2009], Harrell's phantasmal media [Harrell, 2009], and Zhu's AI Hermeneutic Network [Zhu, 2009; Zhu & Harrell, 2009].

11.3 Artificial Intelligence, Cognitive Science and Stream of Consciousness

"Stream of consciousness" is a psychological term that William James coined in his 1890 text *The Principles of Psychology* [James, 1890]. The term was later applied to works by various modernist writers such as Dorothy Richardson, James Joyce, Virginia Woolf, and William Faulkner, indicating both their literary techniques and the genre itself. Beyond various formal experiments, stream of consciousness literature reflects a conceptual purpose—to use the internal thoughts as a primary way of depicting fictional characters. As Humphrey puts it in *Stream of Consciousness in the Modern Novel* [Humphrey, 1954], the works under this genre replace the motivation and action of the "external man" with the psychic existence and functioning of the "internal man."

Decades have passed since modernist authors' initial experiments and many works associated with this literary experiment have entered the canon of "high" literature. Their approach to expressing aspects of human subjectivity and pre-speech consciousness nevertheless are still relevant to many recent technologies (AI), theories (e.g., cognitive linguistics), and forms (computational narrative). These younger developments, in their own ways, have taken the modernist writers' steps further in ways described below.

In this section, we call attention to underlying parallels and synergies between stream of consciousness literature, cognitive linguistics, and AI, as the motivation of our computational narrative project, described in the next sections. We believe that concerns of modernist writers regarding inner thought have been reinvigorated in light of these contemporary cognitive scientific developments regarding preconscious conceptualization. As a critical technical practice, our work in algorithmically narrating characters' memory, reverie, and daydreaming (Section 11.4) exemplifies a new literary form that can leverage AI technologies for expressive narrative without being burdened by the former's philosophical baggage or implicit aesthetic dictates.

11.3.1 *Stream of Consciousness Literature and Artificial Intelligence*

Stream of consciousness writing and AI may pose an unlikely match as a subject of comparative analysis. The two fields not only sprouted in different historical periods, but also reside in two separate communities. One was populated in the early twentieth century and is now associated with academic literary analysis more often than being seen as vibrant area for active creative production, whereas the other is a still on-going development in the techno-science sphere that underwent significant self-reevaluation after the so-called "AI-Winter" of the 1970s [Russell &

Norvig, 2002]. Beneath the obvious differences, however, are the similar overarching goals in their respective historical contexts and parallel roles that they both take on in their relationships to contemporaneous concerns.

First, stream of consciousness literature and AI speak to each other through a shared ambition. Humphrey observed that "[t]he attempt to create human consciousness in fiction is a modern attempt to analyze human nature" [Humphrey, 1954]. If stream of consciousness writers sought their answers by portraying humans directly, the AI community pursued theirs by constructing the "other"—machines. AI practitioner Michael Mateas recently echoed that "AI is a way of exploring what it means to be human by building systems" [Mateas, 2002]. These systems, built in attempt to resemble or surpass their human creators, have become our mirrors to reflect upon our identities as humans [Turkle, 1984].

Secondly, both fields rejected behaviorism in their respective historic periods, and turned their attentions to what happens internally in human mental activities as gateways to understanding "human nature." Prior to the turn of the twentieth century, fictional characters were typically represented by their external behaviors. Writers carefully crafted their actions, dialogues, and rational thoughts to create distinctive personas for their stories. What stream of consciousness writers were able to achieve, in comparison, was to create their characters mainly out of their psychological aspects, including their buzzing random thoughts and associative trails.

The scientific community from which AI grew out of in the 1950s, in parallel, was similarly dominated by behaviorism. The paradigm was based on the laws of stimulus-response and declared itself as the only legitimate scientific inquiry. Mental constructs such as knowledge, beliefs, goals and reasoning steps were dismissed as unscientific "folk psychology" [Russell & Norvig, 2002]. Part of AI's contribution was to bring these scientific taboo back to the table by building powerful computational systems based on them. Like Newell and Simon's 1957 *General Problem Solver* [Newell et al, 1959], many research efforts have been poured into modeling human cognitive capabilities, including reasoning, planning, and learning.

11.3.2 *Stream of Consciousness Literature and Cognitive Linguistics*

The pre-speech level of thought that was neglected by the AI community has been scrutinized again recently in a new field closely built, in part, upon AI: cognitive science. To contemporary cognitive linguists, such as Gilles Fauconnier, George Lakoff, Mark Johnson, and Mark Turner, this neglected land of consciousness holds the basis for our basic conceptual, and even literary thought [Fauconnier, 1985, Fauconnier & Turner, 2002, Lakoff & Johnson, 1980].

Language is only the tip of a spectacular cognitive iceberg, and when we engage in any language activity, be it mundane or artistically creative, we draw unconsciously on vast cognitive resources, call up innumerable models and frames, set up multiple connections, coordinate large arrays of information, and engage in creative mappings, transfers, and elaborations. Beneath the tip of this iceberg is a wide range of cognitive phenomena that Fauconnier calls "backstage cognition," defined as "the intricate mental work of interpretation and inference that takes place outside of consciousness" [Fauconnier, 2001; Fauconnier & Turner, 2002]. Thus, we could say that cognitive linguists cite phenomena that are even below the unarticulated thought phenomena explored by stream of consciousness authors—but at a level that still addresses conceptualization as opposed to perception, motor-action, or other pre-conscious cognitive phenomena.

11.3.3 *Benefits and Challenges*

In the course of working with these very different traditions and methods, we encountered numerous challenges and also compensated by many new perspectives and insights. Many obstacles are a direct result of the clashing differences between the Turing-land and the Duchamp-land. Although both communities were interested in cognitive phenomena, stream of consciousness writers and AI practitioners emphasize different stages of human consciousness. The term "consciousness" from the vantage point of modernist writers referred to "the whole area of mental processes, including especially the pre-speech levels" [Humphrey, 1954].

This was based on James' original psychological theory, in which "memories, thoughts, and feelings exist outside the primary consciousness" and, further, that they appeared, not as a chain, but as a stream, a flow [James, 1890]. Early AI, on the other hand, regarded human rationality as the key to problem solving. Early practitioners in the field relied upon the rational and stable operations of our cognitive processes at the cost of the addressing the roles of the body, affect, and the uncontrollable stream of thoughts unmediated by logic and rationality.

Another conflict is due to the two cultures' opposite values of specificity and generalizability. Modernist writers such as Virginia Woolf believed that the important subject for an artist to express was her private and subjective vision of reality. Woolf's characters all embodied her belief in the individual's constant search of meaning and identification [Humphrey, 1954]. This individualistic approach contrasts strongly with AI's focus on generalizability, in which individual differences are often sacrificed for regularity and scalability. We reconcile these two stances by distancing our work from an attempt to reduce mental activities to uniform formal algorithmic processes. Instead, our project utilizes scientific computational methods, including logical/mathematical formalization, as a way to express our human search for meaning. Formal representations are no more or less meaningful than any other human forms of symbol making; their benefits of precision and computational implementability come at a cost of subjective interpretability.

In the meanwhile, forging the bond across the cultural gap provides unique opportunities. It may be argued that one of the reasons that early AI largely confined itself to the territory of rationality is the extreme difficulty that the field ran into in its attempt to model common sense and contextual reasoning explicitly. These powerful, but for the most part invisible, operations are seen within the field of cognitive linguistics to be partially observable in the structure of our linguistic creations. The insights posed by the cognitive linguistics enterprise and the expressive needs invigorated by our interdisciplinary approach offer the opportunity to revisit some of the compromises that AI made in its early stage.

11.4 Memory, Reverie Machine

The *Memory, Reverie Machine* (MRM) project reflects our approach for invigorating the cultural concerns of depicting internal reflection and imaginative subconscious and for applying cognitive science and AI techniques for expressive purposes. MRM is an interactive narrative system that generates different stories based on user input. The main character of these story is a robot called Ales who gradually discovers himself. In the beginning of the story, Ales functions more like an avatar and is controlled completely by the user. As the story progresses, however, he starts to recall memories triggered by the artifacts and other characters he interacts with. These memories construct Ales' beliefs and desires and determines his emotional states. In the later stage of the story, if the user commands him to perform actions contradictory to his beliefs and desires, Ales may ignore it. Even if he does it eventually, he will do it very reluctantly. At the end of the story, Ales may gain its full autonomy by acting on its own completely or getting lost in his internal world forever.

The sample output in Fig. 11.1 illustrates one iteration of user interaction with the system, containing content from both the main (objective) story world and Ales' internal memory world. Depending on user input, the system will generate different text with different emotional tone in ways that we will discuss below. For the rest of the section, we describe the influences of forbearers in conjoining concerns of computing and literature, and foundational work by one of the authors for our current project. We then highlight the model proposed by MRM using an example comprised of actual system output.

```
the color of the kitchen door today should be [yellow/blue]
-> blue
ales said to himself, adjusting his visual scan system.
the room where he had his first encounter of tune-up and oil
change
had similar depressing entryways.
the oil change left a sickly feeling in his gut;
he would rust like the tin man before enduring another.
it amazes him that no one else complained about these things;
they were too busy with the kitchen;
he should probably start his daily routine now.
```

Fig. 11.1. Sample output from *Memory, Reverie Machine*. User input is indicated by "->".

11.4.1 *Literary Antecedents*

The goal of stirring human imagination through literary works that are different on each reading is not new. It often involves explorations of creative integrations of mathematical and/or algorithmic concepts and writers' meaningful expressions. An early work is Raymond Queneau's 1961 *Cent Mille Milliards de Poémes* (*One Hundred Thousand Billion Poems*), originally published as a set of ten sonnets with interchangeable lines [Queneau, 1961]. It explores the idea of writing as a combinatorial exploration of possibilities. Another member of the experimental literary group Oulipo, Italo Calvino, in his essay/lecture *Cybernetics and Ghosts* [Calvino, 1982], claimed that writing was a combinatorial game and "the operations of narrative, like those of mathematics, cannot differ all that much from one people to another, but what can be constructed on the basis of these elementary processes can present unlimited combinations, permutations, and transformations." In Calvino's novels such as "If on a winter's night a traveler" there was a strong sense of narrative coherence, yet Calvino also carefully explicated the algorithmic generation of the novel's form [Calvino, 1995].

The introduction of AI to electronic literary works laid a foundation for the natural integration of AI, cognitive science, and literary concerns. One of the first computer story generation systems, Meehan's Tale-Spin [Meehan, 1981] produced simple animal fables, with the goal of exploring the creative potential of viewing narrative generation as a planning problem, in which agents select appropriate actions, solve problems within a simple simulated world, and output logs of their actions. A more recent example is Selmer Bringsjord and David Ferrucci's 2000 BRUTUS system, which aims to explore formalizations for generating stories about betrayal, with the goal of being "interesting" to human readers.

It is worthwhile to call attention to the different approaches in these two computer science-based systems. While Tale-Spin directly exposed a reader to the output of a planning algorithm, BRUTUS used rich textual descriptions to accentuate the narrative outcome, an approach native to the Duchamp-land. In fact, BRUTUS's extensive amount of pre-authored output raised many questions, especially among the computer science community, whether the system actually authored the text. This criticism is only valid if the system is said to be an autonomous author; in our opinion, there is nothing wrong with human authors creating computationally reconfigurable texts).

In contrast, William Chamberlain and Thomas Etter's dialogue-based program Ractor, and Ractor's book *The Policeman's Beard is Half Constructed* [Racter, 1984], used syntactic text manipulation to support conversation with users having text input and poetic output. This was not intended as scientific research, but rather as entertainment, with humorous and clever output. As Charles Hartman [1996] stated, it is better not to ask "whether a poet or a computer writes the poem, but what kinds of collaboration might be interesting."

11.4.2 *Conceptual Blending and the GRIOT System*

The cognitive semantics theory of conceptual blending and the GRIOT system are the foundations of the MRM project. In contrast to the notion of computational generativity, the human capacity to generate concepts and metaphors has been explored by cognitive scientists as the root of our literary mental processes. Conceptual blending theory, building upon Gilles Fauconnier's mental spaces theory [Fauconnier, 1985] and elaborating insights from metaphor theory [Lakoff & Johnson, 1980, Turner, 1996], describes the means by which concepts are integrated. Simple examples of blending in natural language are reflected in the mental processes triggered by words like "houseboat" and "roadkill," and phrases like "artificial life" and "computer virus." In short, the theory describes how we arrive at new concepts through blending partial and temporary pieces of information. Most importantly, the theory proposes that conceptual blending processes occur uniformly in pre-conscious in everyday thought and in more complex abstract thought such as in literary arts or rhetoric [Fauconnier & Turner, 2002].

The GRIOT system, a platform for implementing phantasmal media in the form of generative and interactive multimedia works, is the foundation of MRM both in terms of technical implementation and our approach to computational narrative [Harrell, 2006; 2007a]. This subsection, adapted from the abstract of [Harrell, 2007b], serves as a high level overview of this perspective, which emphasizes computational narrative works with the following characteristics: generative content, semantics-based interaction, reconfigurable narrative structure, and strong cognitive and socio-cultural grounding. A system that can dynamically compose media elements (such as procedural computer graphics, digital video, or text) to result in new media elements can be said to generate content.

GRIOT's generativity is enabled by blending-based concept generation as described above. It uses Joseph Goguen's theory of algebraic semiotic approach from computer science to formalize key aspects of conceptual blending theory [Goguen, 1998]. Technical details can be found in [Goguen & Harrell, 2004]. Semantics-based interaction means here that

1. media elements are structured according to the formalized meaning of their content, and
2. user interaction can affect content of a computational narrative in a way that produces new output that is "meaningfully" constrained by the system's author.

More specifically, "meaning" in GRIOT indicates that the author has provided formal descriptions of domains and concepts to either annotate and select or generate media elements and subjective authorial intent.

Meaning can also be reconfigured at the level of narrative discourse. The formal structure of a computational narrative can be dynamically restructured, either according to user interaction, or upon execution of the system as in the case of narrative generation. Discourse structuring is accomplished using an automaton that allows an author to create grammars for narratives with repeating and nested discourse elements, and that accept and process user input. Appropriate discourse structuring helps to maintain causal coherence between generated blends. Strong cognitive and socio-cultural grounding here implies that meaning is considered to be contextual, dynamic, and embodied. The formalizations used derive from, and respect, cognitive linguistics theories with such notions of meaning. Using semantically based approach, a cultural producer (someone producing expressive works, though potentially not self-defined as an "artist") can implement a range of culturally specific or experimental narrative structures. In the subsection below we describe our new work that arises from the historical context presented above and the theoretical and creative framework just described.

The goal with GRIOT is quite different from passing a type of Turing test for autonomous creative competence, as described earlier this section. It is designed to provide a technical framework for humans to provide rich content; narrative systems created with GRIOT are meant as cultural products themselves (as opposed to instances of output of such poetic systems). The GRIOT system utilizes models for cognitive science, informed by the cognitive linguistics enterprise's skepticism of regarding the possibility of a formal account of human thought and language such as in [Lakoff & Johnson, 1999], toward expressive ends that are often literary.

11.4.3 *Framework for Memory, Reverie Machine*

So far, we have situated our work in a historical context where stream of consciousness literature, artificial intelligence discourse, and cognitive science research complement each other, as well as the technical framework of this project. In this subsection, we present MRM, a text-based computational narrative project through one of our early results (illustrated in Fig. 11.2 below). For this work it is important to distinguish between the project's various levels of technological and expressive investigation and production:

1. the system as an abstract model for how computational narratives can be made generative, extensible, and reconfigurable,
2. the system that generates the story,
3. the narration techniques developed,
4. the story content, and
5. each instance of output.

The emphases in this chapter are upon levels 1 and 2, which comprise our technological framework and secondarily on level 3, the narrative techniques to depict inner thoughts, and level 4 the self-reflexive subject matter. Level 3 is influenced by Virginia Woolf's stream of consciousness novel *Mrs. Dalloway* [Woolf, 2002 (1925)]. Below we highlight particular aspects of the system relevant to the algorithmic narration of inner-thought; complete technical information on GRIOT and of MRM can be found in [Zhu & Harrell, 2008].

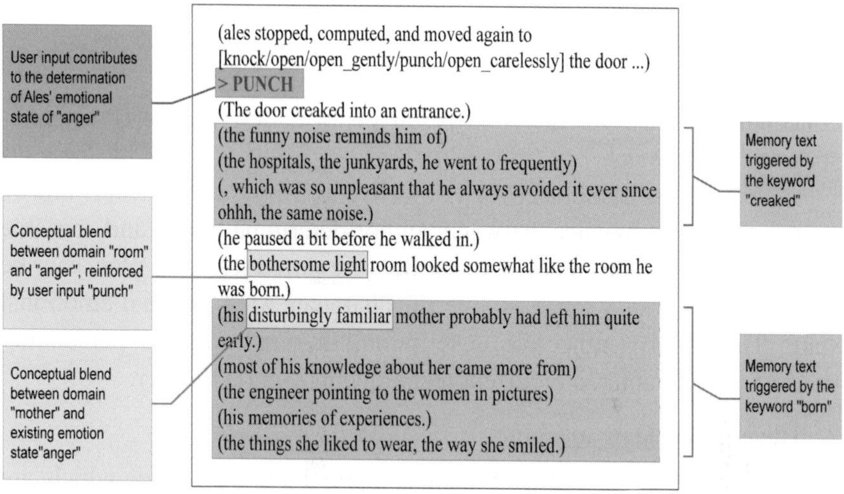

Fig. 11.2. An example of output from Memory, Reverie Machine.

1. Dynamic narration of affect using the Alloy conceptual blending algorithm

Computationally, our system draws upon the GRIOT framework (Section 11.4.2), whose primary generative component is the Alloy algorithm. When modestly applied in MRM, Alloy generates blends involving connecting the main character's current experiences of events, objects, and other characters with affective concepts determined by his current emotional state (see next subsection). In the example in Fig. 11.2, logical axioms selected from an ontology (semantically structured database) describing the concept "room" are blended with axioms describing affective concept "anger," resulting a "bothersome light" room.

The Alloy algorithm uses a set of formal optimality criteria to determine the most common-sense manner in which the concepts should be integrated [Harrell, 2007b]. The result is a blended axiom or set of axioms that is then mapped to natural language output. For example, the description of the "door" in the "anger" state may range from "distasteful wood-colored" to "irritatingly sturdy" or more depending on the concepts being blended. Since blending refers to the conceptual integration of multiple concepts, it is important to be clear that blending is not the mere concatenation of words to form compound phrases. In this case, compound phrases, some of the simplest indicators of conceptual blends, are the final result of an underlying process that is semantic, not lexical.

Constructing blends between objective and affective concepts allows us to achieve a balance between author-determined plot and variable theme or emotional tenor. An artifact required by the plot can be depicted in various ways based on the character Ales' internal emotional state. The highly subjective description, in turn, portrays personality traits of the character, a recurrent technique in *Mrs. Dalloway*.

2. The emotion state machine

Actions taken by a character in a computational narrative, which are usually (but not exclusively) selected by a user, can guide building up of a profile that describes user's preferences, history of actions, and analysis of trends in those actions. A quite simple, but effective, way to do this is tracking tokens representing the emotional state of a character based upon actions that the character has taken. MRM allows user to directly influence the emotion state of Ales, and hence the selection of affective concepts for blends. She may choose among an array of pre-defined actions, such as seeing objects as "red," "yellow," "blue" or another color in the fictional robot character's optic sensors, each connecting to a particular emotion. A keyword "red," for instance, may trigger an affective concept "anger." These emotional mappings are designed aesthetically by the authors to achieve narrative effects, not as an attempt at cognitively modeling emotion using computers as in multiple traditional AI projects.

A successful interactive narrative, however, requires a careful balance between the user's agency and author's intention. In our system, user's impact on the character's emotion is moderated by the emotion state machine component for the sake of narrative consistency. The state machine records Ales' current emotion based on the entire history of user input, instead of the most immediate one. It guarantees that changes of Ales' emotions will be gradual, even if user input oscillates between opposite emotions.

3. Memory structuring and retrieval

The GRIOT system is not limited to producing narrative discourse; indeed it has been used for various forms of poetry [Goguen & Harrell, 2004, Harrell, 2007a], for the interactive, generative composition of animated imagery [Chow & Harrell, 2008] and digital video, photography, and illustration.[3] In the case of MRM, we seek to make output coherently extensible at runtime. For this project we allow the narrative to be punctuated with episodic remembered events and longer reveries of remembered experience. Again, this is not meant as an experiment in cognitive modeling or advanced algorithmic design, our goal is to demonstrate discourse that is meaningfully reconfigurable to serve an author's expressive goals in dialogue with a user's selected actions. In MRM, each memory is annotated based on its subject matter and is retrieved when at least one subject item appears in the story line. In the example in Fig. 11.2, Ales' unpleasant memory of hospitals and junkyards is triggered by the opening of a door through the mutual subject of a certain sound. The system also keeps track of the emotional tone of each memory and selects a memory only if it does not clash with the current emotion state. The example in Fig. 11.1 illustrates this feature.

[3] One of these artworks is titled "Authoring the Living Liberia Fabric: A Generative and Interactive Narrative for Peace, Truth, and Reconciliation," which is shown in the juried exhibition at the 2010 Electronic Literature Organization Conference in Providence, RI. (Author: D. Fox Harrell, Michael Best, Hank Blumenthal, Ayoka Chenzira, Christopher Gonzalez, Andrew Roberts, Natasha Powell, Deji Fajebe, Jason Lee, Paul O'Neil, and Arjun Tomar.)

MRM represents an approach that is different from Duchamp-land or Turing-land, explained in Section 11.2, for neither culture is seen as subordinate to the other. We see our approach closely related to "bootstrapping," a terminology in computing that describes how simple programs can build up larger ones. In our project, the exchange between the two cultures strengthens both: our new insights in cognitive science theory and discoveries in our AI techniques help us to further steer and articulate our expressive goal; renewed aesthetic needs guide the next iteration of algorithmic exploration. In our case, this tightly coupled feedback loop led to tremendous improvements in both aspects.

11.5 Conclusion

Digital technology has no doubt created new borderlands where the arts and sciences intersect: English scholars apply digital statistical tools to hundreds of books and identify patterns and trends beyond the human scale; [4] dancers use computer vision technology to mediate their performances [Nahrstedt et al, 2008]; and computer scientists use social theory to inform their multi-agent AI systems [McCoy & Mateas, 2009], and more. Yet, despite the growing interest and knowledge between the two cultures, however, Snow's cultural divide still persists.

We believe these interdisciplinary collaborations can be further extended if we can take advantage of each discipline's strength. More importantly, one can use the other culture as a mirror to make visible and reflect upon the hidden value system and ideologies in our practices. In the case of our own interactive narrative project Memory, Reverie Machine, we intend to forge a unique bond: engaging stream of consciousness literature and critically examining subjective experiences of AI technologies without taking a totalizing, modernist stance toward literary production, and engaging cognitive science and AI as a critical technical practice where both the methods used and interactive output are seen as expressive.

[4] This approach, for example, is illustrated by the various works in the "Computational Approaches to Textual Variation in Medieval Literature" Panel in the 2010 Digital Humanities Conference, held in London, UK.

While confronting the other culture, Agre [1997b] stated the importance of "[m]aintaining constructive engagement with researchers whose substantive commitments I found wildly mistaken. It is tempting to start explaining the problems with these commitments in an alien disciplinary voice, invoking phenomenology or dialectics as an exogenous authority, but it is essentially destructive." It is with this constructive approach—looking not only at the wildly mistaken, but also the wildly inspiring—that we, along with our peer practitioners in diverse fields explore the borderland between art and science. We hope that others might engage our own work as at least "wildly mistaken" in interesting, useful, and productive ways. More hopefully, it can be seen as a productively "wild" reconciliation of disciplinary values and expressive practice as we forge ahead in our hybrid research/art.

References

Agre, P. E. [1997a] *Computation and Human Experience* (Cambridge University Press, Cambridge).

Agre, P. E. [1997b] "Toward a critical technical practice: Lessons learned in trying to reform AI," in *Social Science, Technical Systems, and Cooperative Work: Beyond the Great Divide*, eds. Bowker, G. C., Star, S. L., Turner, W. & Gasser, L. (Lawrence Erlbaum Associates, Mahwah, NJ) pp. 131-158.

Bogost, I. [2006] *Unit Operations: An Approach to Videogame Criticism* (MIT Press, Cambridge, MA).

Calvino, I. [1982] *The Uses of Literature* (Harcourt Brace, San Diego, CA).

Calvino, I. [1995] "How I wrote one of my books," in *Oulipo Laboratory: Texts from the Bibliotheque Oulipienne*, ed. Queneau, R. (Atlas, London).

Chow, K. K. N. & Harrell, D. F. [2008] "Generative visual renku: Linked poetry generation with the GRIOT system," *Hyperrhiz: New Media Cultures* **6** (www.hyperrhiz.net/hyperrhiz06/19-essays/79-generative-visual-renku).

Edwards, P. N. [1996] *The Closed World: Computers and the Politics of Discourse in Cold War America* (MIT Press, Cambridge, MA).

Fauconnier, G. [1985] *Mental Spaces: Aspects of Meaning Construction in Natural Language* (MIT Press/Bradford Books, Cambridge, MA).

Fauconnier, G. [2001] "Conceptual blending and analogy," in *The Analogical Mind: Perspectives from Cognitive Science*, eds. Gentner, D., Holyoak, K. J. & Kokino, B. N. (MIT Press, Cambridge, MA).

Fauconnier, G. & Turner, M. [2002] *The Way We Think: Conceptual Blending and the Mind's Hidden Complexities* (Basic Books, New York).

Goguen, J. [1998] "An introduction to algebraic semiotics, with applications to user interface design," in *Computation for Metaphors, Analogy, and Agents*, ed. Nehaniv, C. L. (Springer, Berlin) pp. 242-291.

Goguen, J. & Harrell, D. F. [2004] "Style as a choice of blending principles," in *Style and Meaning in Language, Art, Music and Design: Papers from the AAAI Fall Symposium*, eds. Dubnov, S., Argamon, S. & Jupp, J. (AAAI Press, Menlo Park, CA) pp. 49-56.

Harrell, D. F. [2006] "Walking blues changes undersea: Imaginative narrative in interactive poetry generation with the GRIOT system," in *Computational Aesthetics: Artificial Intelligence Approaches to Happiness and Beauty*, eds. Liu, H. & Mihalcea, R. (AAAI Press, Menlo Park, CA) pp. 61-69.

Harrell, D. F. [2007a] "GRIOT's tales of Haints and Seraphs: A computational narrative generation system," in *Second Person: Role-Playing and Story in Games and Playable Media*, eds. Wardrip-Fruin, N. & Harrigan, P. (MIT Press, Cambridge, MA).

Harrell, D. F. [2007b] *Theory and Technology for Computational Narrative: An Approach to Generative and Interactive Narrative with Bases in Algebraic Semiotics and Cognitive Linguistics* (Ph.D. Dissertation, University of California, San Diego, CA).

Harrell, D. F. [2009] "Toward a theory of phantasmal media: An imaginative cognition and computation-based approach to digital media," *CTHEORY*, Sept. 2 (http://www.ctheory.net/articles.aspx?id=610).

Harrell, D. F. [2010] "Toward a theory of critical computing: The case of social identity representation in digital media applications," *CTHEORY*, May 13 (http://www.ctheory.net/articles.aspx?id=641).

Hartman, C. O. [1996] *Virtual Muse: Experiments in Computer Poetry* (Wesleyan, Hanover, CT).

Humphrey, R. [1954] *Stream of Consciousness in the Modern Novel* (University of California Press, Berkeley).

James, W. [1890] *The Principles of Psychology* (Henry Holt, New York).

Lakoff, G. & Johnson, M. [1980] *Metaphors We Live By* (University of Chicago Press, Chicago).

Lakoff, G. & Johnson, M. [1999] *Philosophy in the Flesh: The Embodied Mind and Its Challenge to Western Thought* (MIT Press, Cambridge, MA).

Lam, L. [2008] "Science Matter: A unified perspective," in *Science Matters: Humanities as Complex Systems*, eds. Burguete, M. & Lam, L. (World Scientific, Singapore) pp. 1-38.

Manovich, L. [1996] "The death of computer art," *Rhizome* (http://www.manovich.net/text/death.html, Oct. 6, 2010).

Mateas, M. [2001] "Expressive AI: A hybrid art and science practice," *Leonardo* **34**(2), 147-153.

Mateas, M. [2002] *Interactive Drama, Art, and Artificial Intelligence*. (Ph.D. dissertation, Carnegie Mellon University, Pittsburgh).

McCoy, J. & Mateas, M. [2009] "The computation of self in everyday life: A dramaturgical approach for socially competent agents," in *Intelligent Narrative Technologies II*, eds. Louchart, S., Mehta, M. & Roberts, D. L. (AAAI Press, Menlo Park, CA) pp. 75-82.

Meehan, J. [1981] "TALE-SPIN," in *Inside Computer Understanding: Five Programs Plus Miniatures,* ed. Riesbeck, C. K. (Lawrence Erlbaum Associates, New Haven, CT).

Nahrstedt, K., Bajcsy, R., Wymore, L., Sheppart, R. & Mezur, K. [2008] "Computational model of human creativity in dance choreography," in *Creative Intelligent Systems: Papers from the 2008 AAAI Spring Symposium*, eds. Ventura, D., Maher, M. L. & Colton, S. (AAAI Press, Menlo Park, CA) pp. 53-60.

Newell, A., Shaw, J. C. & Simon, H. A. [1959] "Report on a general problem-solving program," in *Proceedings of the International Conference on Information Processing*, 1959, pp. 256-264.

Penny, S. [2004] "Representation, enaction, and the ethics of simulation," in *First Person*, eds. Wardrip-Fruin, N. & Harrigan, P. (MIT Press, Cambridge, MA).

Penny, S. [2007] "Experience and abstraction: The arts and the logic of machines," *The Fiberculture Journal* **11** (http://journal.fibreculture.org/issue11/issue11_penny.html).

Queneau, R. [1961] *Cent Mille Milliards de Poémes* (Gallimard, Paris, France).

Racter [1984] *The Policeman's Beard Is Half Constructed* (Warner Books, New York).

Russell, S. & Norvig, P. [2002] *Artificial Intelligence: A Modern Approach* (Prentice Hall, Upper Saddle River, NJ).

Sengers, P. [1998] *Anti-Boxology: Agent Design in Cultural Context* (Ph.D. Dissertation, Carnegie Mellon University, Pittsburgh).

Sengers, P. et al [2004] "Culturally embedded computing," *Pervasive Computing* **3**(1), 14-21.

Snow, C. P. [1964] *The Two Cultures* (Menton Books, New York).

Suchman, L. [1987] *Plans and Situated Actions: The Problem of Human-Machine Communication* (Cambridge University Press, Cambridge).

Turkle, S. [1984] *The Second Self: Computers and the Human Spirit* (Simon & Schuster, New York).

Turkle, S. [2004] "Whither psychoanalysis in computer culture?" *Psychoanalytic Psychology* **21**, 16-30.

Turner, M. [1996] *The Literary Mind: The Origins of Thought and Language* (Oxford University Press, Oxford).

Wardrip-Fruin, N. [2009] *Expressive Processing: Digital Fictions, Computer Games, and Software Studies* (MIT Press, Cambridge, MA).

Weizenbaum, J. [1976] *Computer Power and Human reason* (Freeman, New York).

Winograd, T. & Flores, F. [1986] *Understanding Computers and Cognition* (Ablex, New York).

Woolf, V. [2002 (1925)] *Mrs. Dalloway* (Harcourt, New York).

Zhu, J. [2009] *Intentional Systems and the Artificial Intelligence Hermeneutic Network: Agency and Intentionality in Expressive Computational Systems* (Ph.D. Dissertation, Georgia Institute of Technology, Atlanta, GA).

Zhu, J. & Harrell, D. F. [2008] "Daydreaming with intention: Scalable blending-based imagining and agency in generative interactive narrative," in *Creative Intelligent Systems: Papers from the 2008 AAAI Spring Symposium*, eds. Ventura, D., Maher, M. L. & Colton, S. (AAAI Press, Menlo Park, CA) pp. 156-162.

Zhu, J., & Harrell, D. F. [2009] "The artificial intelligence (AI) hermeneutic network: A new approach to analysis and design of intentional systems," in *Proceedings of the 2009 Digital Humanities Conference*, 2009, pp. 301-304.

12

ChemArt and BioArt: Art-Science Interactions

Maria Burguete

The development of science is a time evolving process; so is Art. However Art began earlier and so far they still do not have a methodology the same way science has. Why does this happen? For a scientist truth results from the scientific methodology; for an artist truth comes from revelation plus inspiration. The structural unity of art can be seen as an expression, in the sense of an iconographic translation of the cosmic universe. Art deals with symbolism based on a platform that can move frontiers by overcoming the limits of common perception. Art is a place where new worlds are invented, and also where art itself is continuously being created. *ChemArt* refers to those artworks that use images or concepts from chemistry as the central theme, and occasionally, chemicals as a medium; similarly for *BioArt*, except that chemicals are replaced by biological materials. Considering ChemArt as the interaction between Chemistry and Art, we can say that flexibility of nature enhanced the emergence of both chemistry and art: Chemistry allows us to understand diversity and beauty of nature at the molecular level, and awakens our curiosity about its "modus operandi"; Art, through the interaction of human beings with the environment, induces an expressive reaction by contemporary artists. ChemArt naturally leads us to BioArt, the same way that yesterday's chemistry leads us to the modern chemistry of living organisms or biochemistry. BioArt is the new challenge the contemporary world is facing now: a crossroad between artistic culture and scientific culture. A major question dealing with science and art is concerned with the incorporation of biological and biomedical tools in the performance of artistic works, and the use of living tissues as an *art medium.* This new approach strongly appeals to a straight cooperation and collaboration among artistic production and scientific practices. The ultimate aim of science is to look for a single unifying "theory of everything." Art does the same from an iconographic point of view, and includes spirituality.

12.1 Introduction

Deep analogies are displayed between the working processes of experimental scientists and visual artists who detect new fragments of reality through the perception of their own data. In both processes similar roles are played by active perception and imagination, therefore emerging new areas of knowledge, known as *ChemArt* and *BioArt* (Fig. 12.1). Here, ChemArt refers to those artworks that use images or concepts from chemistry as the central theme, and occasionally, chemicals as a medium; similarly for BioArt, except that chemicals are replaced by biological materials.

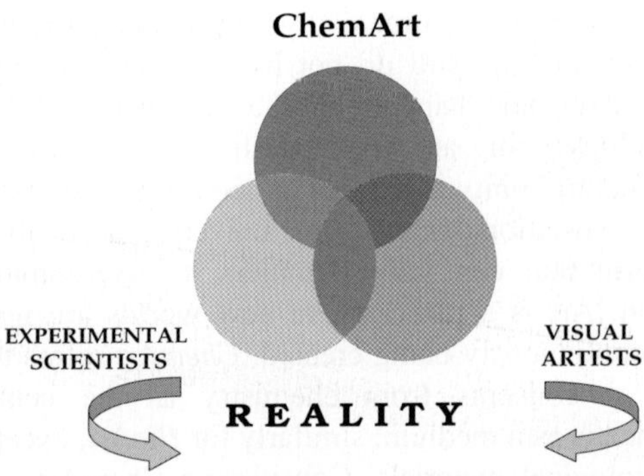

Fig. 12.1. ChemArt (top circle) results from the interaction between science (left circle) and art (right circle). Experimental scientists and visual artists can reach reality through science and art, respectively.

Let us define *reality* as the group of permanent relationships to which our minds can access while processing their signs or information. Both art and science allow us to grasp reality. Whether the end result is a work of science (Fig. 12.2) or a work of art (Fig. 12.3), the act of creating has many similarities. Creative work can also be analyzed from a philosophical point of view or from that of neuroscience.

Scientific concepts themselves can also be used to describe the understanding of some aspects of art: *artistic science*. And these concepts can also influence or inspire artists who use science as the subject matter or basis of their work: *scientific art*.

Fig. 12.2. Scientists working in 1866. *Upper*: Liebig's Laboratory at Giessen University. *Lower*: Scientists in the laboratory. [Pictures from Deutsches Museum Archives.]

Fig. 12.3 Artist working: Gustave Courbet in his studio at 19th century.

When scientists want to describe or understand a reality that is beyond what we can see with our eyes, e.g., the infinitely small, they cannot refer to our usual intuition, which is formed during our experience with objects on our own scale. Scientists have to acquire a new intuition, to discover or invent new concepts or find new ways of thinking that perhaps are not specific to science but that science has brought to light.

New scientific concepts have the capacity to enlarge the imagination and the artistic vocabulary of an artist. A good example of such a new concept is that of the "wave-particle" dualism: sometimes behaving like a traditional wave and sometimes like a traditional particle but is different from both. A good movie that illustrates this "wave-particle" concept is *What the Bleep Do We Know?* (2004).[1]

Let us consider ChemArt as a case study where the interactions of chemistry and art have been studied. Two major items will be presented here:

1. *Artistic Science* where we will see how science can be done by the use of artistic analogies in chemical research and education—bridging gaps between the familiar and the unknown. Therefore, we have artistic science when scientific concepts such as geometry, packing, rotation and molecular models can be used to describe some aspects of art. Some examples will be presented to illustrate these concepts such as drawings taken from the remarkable picture by Edgar Degas, *Prima Ballerina* (1876) that can be seen at Louvre Museum in Paris.
2. *Scientific Art* where some artists inspired by scientific concepts or biological forms, create their work of art. Here we will see the role of scientific concepts in art by illustrative examples given with the artworks performed by Nicole Aggagio and others.

ChemArt naturally leads to BioArt which emerges whenever there is incorporation of biological and medical tools in the performance of artistic works as well as the use of living material (e.g., DNA, proteins,

[1] It is about human conscience and daily "reality" experienced by a woman (acted by Marlee Matlin).

cells and organisms) as art medium. This will be illustrated by the artworks of Marta de Menezes [2004], an artist that has explored several different medical and biological techniques to create her artworks.

12.2 Science and Art

Science provides us with insight about our world and how we relate to it. This insight is based not only on scientific results or concepts but also on the way science is made using a scientific methodology (Fig. 12.2), the characteristics of which are: regularity, generality and a critical methodology [Careri, 1994].

The artist in the studio (Fig. 12.3) reveals a different picture of the world that surrounds him and here we witness the diversity, uniqueness and individuality characterizing a work of art. I would like to present a citation from Georges Braque taken from [Charbonier, 1980]:

> One should start from a leading idea. But in the working process, the picture affirms itself. When the picture is free from the idea, the picture takes on a life of its own. In a picture what matters is the unpredictable, this is what remains. A casual fact, in my opinion, brings life and spontaneity to the picture.

12.3 ChemArt: A Case Study

Chemistry is a creative science and the fact that chemistry creates a major part of its objects by synthesis prompted French chemist Marcelin Berthelot (1827-1907) to compare chemistry to the creative arts a long time ago. It is said that Chemistry is a harder science than Biology but a softer one than Physics, in reference to the amount of mathematics and exactness involved [Burguete, 2008].

When we hear sentences like "Abstract art is cold. And so is science." we should notice that when we look at an abstract art picture or a chemical mechanism we could develop both emotional and cognitive reactions, because both art and science share the same problem of perception. In fact, as Roald Hoffmann [2009] said,

> When the figurative, natural or personal are pushed aside, the cognitive is emphasized. And thinking seems less warm. Oh, beauty comes back, no way of keeping its subversive pleasures out of the soul. But the initial impression of a shift in emphasis from the emotional to the cognitive is... coldness.

Opinions may differ as to whether there is any art in chemistry but it has been noted that analogies taken from the arts and humanities have proven very fertile in both chemical research and education [Prelog, 1976; Hoffmann, 2009].

12.3.1 *Artistic Science*

The use of simple analogies in research and education facilitates the cognitive process by establishing connections between the familiar and the unknown. A sample of examples includes works by Degas.

Symmetry has proven particularly useful as a bridging tool in relating known and unknown concepts, ideas and observations. Close–packing is an important concept in crystallography. Analogies were used to stress the point, and they came in the form of simple but powerful statements. Two such analogies are quoted here [Hargittai & Hargittai, 1994]: In 1982 A. Kitaigorodskii said, "The molecule also has a body. When you strike it the molecule feels hurt all over"; in 1985 P. Diehl said, "Molecules are like people. Either they are free and this is marvelous, or they are oriented, but then they are not without deformation."

Degas' drawings (Fig. 12.4) have been used to make it easier to grasp the idea of projection representation for rotational isomers. This analogy first appears in the *Journal of Chemical Education* in 1963 (Vol. 60, p. 94), and one of the dancers appeared on the cover. Note that Degas' dancers provide only an external similarity.

Another good example of artistic science is a computer-generated stereo slide pairs of Soccer Ball Molecule, fullerene molecule (C_{60}) composed of 60 carbon atoms linked together in a ball made by Kenneth Snelson in 1994 (Fig. 12.5). In this picture the double images of each of these stereo slide pairs can be seen as apparently three-dimensional through "free fusion" [Snelson, 1994; Soto, 1994]. To achieve stereopsis the viewer holds the pair images about 1½ cm from the eyes and gazes at the image without actually focusing on it, almost as if viewing a distant planet. This should reveal not two but three frames: the center one being a combined overlay fusion of the two real pictures. Stereo images allow us to see 3D images while using a 2D picture.

Kenneth Snelson says[2] that his art is concerned with nature in its primary aspect, the patterns of physical forces in 3D spaces. He believes that quantum scientists have been quick to dismiss the visual atom. He argues, "I agree that the uncertainty principle makes sense, but to say therefore that we should no longer ask these questions, that the atom has no structure because we can't see it, is silly. If cosmologists took that attitude, we had never have had any speculations about the big bang or the steady-state universe."

Fig. 12.4. Illustrations of the projection representation of rotational isomerism. *Left*: Staggered conformer; a drawing after Degas' *End of Arabesque*. *Right*: Eclipsed conformer; a drawing after Degas' *Seated Dancer Adjusting Her Shoes*. [Drawings courtesy of Ferenc Lantos.]

---

[2] Private communication (1981).

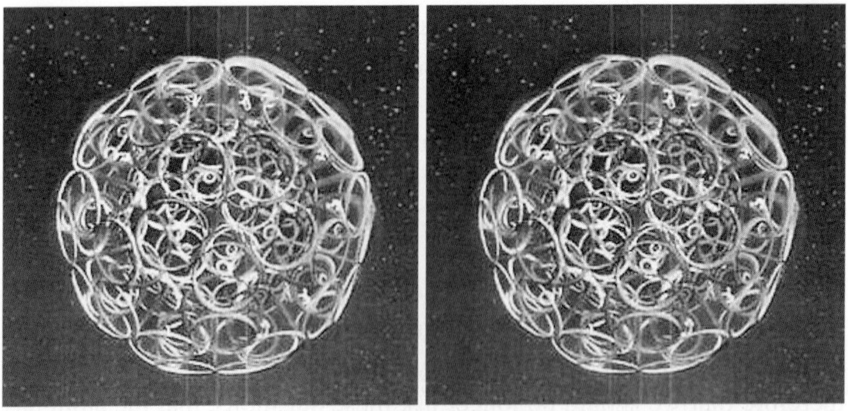

Fig. 12.5. Kenneth Snelson's computer-generated stereo slide pairs, C_{60} Soccer Ball Molecule. The double images of each of these stereo slide pairs can be seen as apparently three-dimensional through "free fusion" (see text).

12.3.2 *Scientific Art*

For a long time, artists have been inspired by biological forms and patterns. The anatomical drawings made by Leonardo da Vinci around 1514 are a good example of such works of art. To illustrate a book that presents a molecular vision of the world, Nicole Agaggio together with Raymond Daudel made paintings inspired by molecules of biochemical interest [Daudel & Agaggio, 1994] which can be seen in Figs. 12.6-12.8.

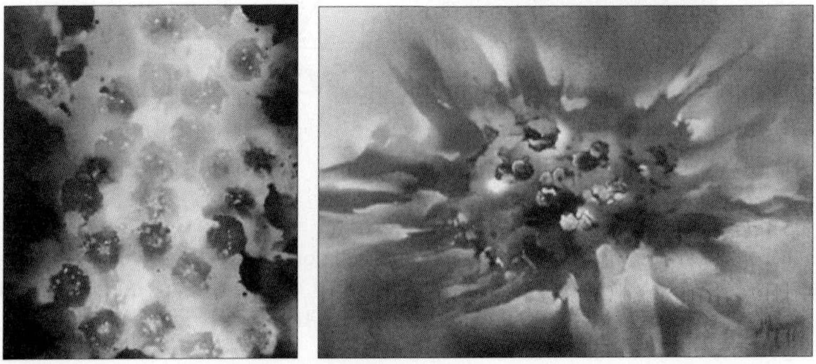

Fig. 12.6. Nicole Agaggio, *Molecular Coupling* (left); *The First Living Molecules* (right); both: oil on linen-faced paper, 85 x 70 cm (1978).

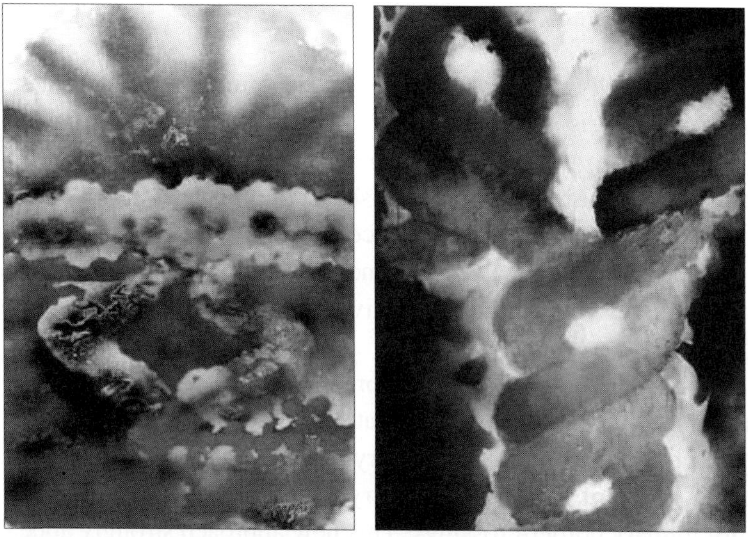

Fig. 12.7. Nicole Agaggio, *Cellular Division* (left); *DNA Helix* (right); both: oil on linen-faced paper, 85 x 70 cm (1978).

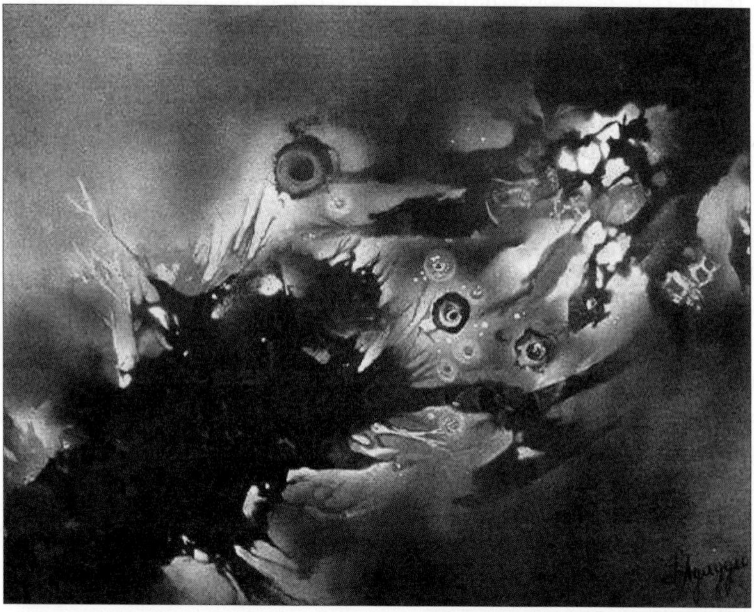

Fig. 12.8. Nicole Agaggio, *Molecules Supporting Thoughts* (1977), oil on linen-faced paper, 85 x 70 cm.

12.4 BioArt

ChemArt naturally leads us to BioArt, the same way that yesterday's chemistry leads us to the modern chemistry of living organisms or biochemistry. BioArt is the new challenge the contemporary world is facing now: a crossroad between artistic culture and scientific culture. A major question dealing with science and art is concerned with the incorporation of biological and biomedical tools in the performance of artistic works as well as the use of living tissues as art medium [Kemp, 2000].

In some cases the scientific laboratory can be used as an *atelier*; in some others the artist's private atelier in a Montmartre loft has metamorphosed now into a laboratory space placed at the crossroads of multiple lines of scientific inquiry, technological innovation and interdisciplinary border-crossings. This new approach strongly appeals to a straight cooperation and collaboration among artistic production and scientific practices.

It all started with SymbioticA,[3] an artistic laboratory dedicated to the research, learning, critique and hands-on engagement of the life sciences. SymbioticA is the first research lab of its kind, in that it enables artists and researchers to engage in wet biology practices in a biological science department. Since then several other artworks have been created using brain fMRI (functional magnetic resonance imaging) techniques to create pictures where the mind can be observed. These artworks have been exhibited at the Biennale of Electronic Arts Perth (BEAP), at Perth, Australia in August 2002 (see www.martademenezes.com).

Marta de Menezes is the artistic director of Ectopy, the Laboratory of Experimental Art at Gulbenkian Institute of Science, Lisbon. Her artistic activity has been developed at scientific research laboratories, exploring new approaches concerning science communication and representation, by using DNA, proteins, cells and organisms as art medium. In her work *Nature?* she had modified the pattern of one wing of *Bicyclus* and *Heliconius* butterflies, respectively (Fig. 12.9), by

[3] At the University of Western Australia, the SymbioticA Research Group is managed by Oron Catts and directed by Miranda Grounds and Stuart Bunt in the School of Anatomy and Human Biology.

interfering with the normal developmental mechanisms of the butterflies.[4] As a consequence, all these butterflies have simultaneously one wing with the natural design and another one with her design. Through this asymmetry, she has tried to emphasize the similarities and differences between the unmanipulated and manipulated, between the natural and the novel natural. She has been trying to express concepts in the butterfly wings that deal with our perception of shapes. By adding, changing or deleting eyespots and color patches it is possible for our imagination to identify shapes and rhythms familiar to our senses.

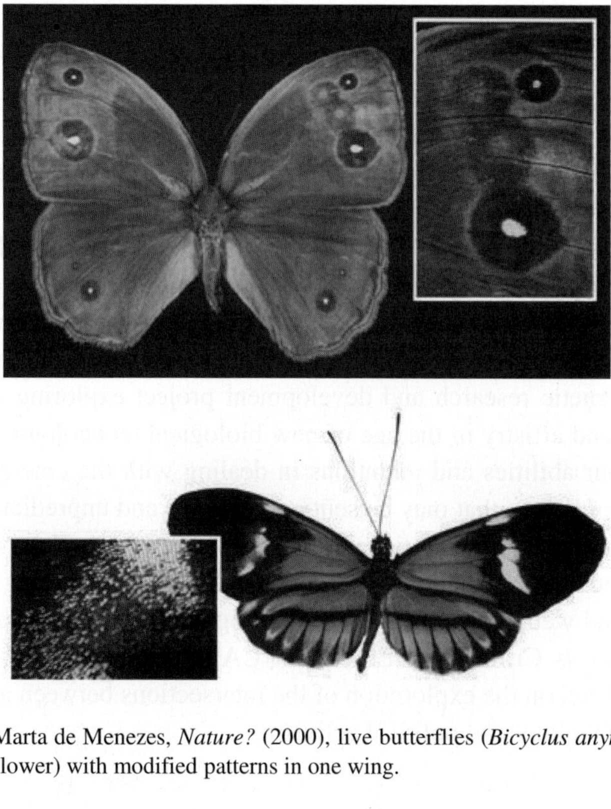

Fig. 12.9. Marta de Menezes, *Nature?* (2000), live butterflies (*Bicyclus anynana*, upper; *Heliconius*, lower) with modified patterns in one wing.

[4] This work was performed during a residency in Paul Brakefield's laboratory at the University of Leiden, the Netherlands. The work was exhibited for the first time in Ars Electronica 2000, Linz, Austria.

The idea for the project came from an article about scientists who modified the wing patterns of butterflies to explore biological questions. She contacted the lead researcher, Paul Brakefield at the University of Leiden, the Netherlands, and ended up spending time at his laboratory, working alongside scientists. "I found it amazing that those butterflies' wing patterns had been modified by scientists, and yet were exclusively made of normal cells, creating a tension between the natural and the artificial," says de Menezes.[5] To my knowledge, it was the first time modified live animals (i.e., not bacteria or plants) for artistic purposes were exhibited in an art venue. Today, biotechnology is her medium of artistic expression. De Menezes is using DNA microarrays, also known as gene chips, to create art about genetic differences between individuals. "I want to make a visual statement about how similar we all are, in spite of the differences we recognize," says the artist.

In Fig. 12.10 we can see another example of a BioArt project, *MEART: The Semi Living Artist*, the robotic arm developed by Phil Gamblen of the SymbioticA Research Group.[6] *MEART* takes the basic components of the brain (isolated neurons) and attaches them to a mechanical body through the mediation of a digital processing engine, to attempt and create an entity that will seemingly evolve, learn and become conditioned to express its growth experiences through "art activity." It is a bio-cybernetic research and development project exploring aspects of creativity and artistry in the age of new biological technologies. *MEART* explores our abilities and intentions in dealing with the emergence of a new class of beings that may be sentient, creative and unpredictable.

Another group internationally well known as a collective of five tactical media practitioners of various specializations including computer graphics and web design, film/video, photography, text art, book art, and performance is Critical Art Ensemble (CAE). Formed in 1987, CAE's focus has been on the exploration of the intersections between art, critical theory, technology and political activism.

---

[5] Private communication (2001).

[6] See detailed description of *MEART* at www.fishand-chips.uwa.edu.au/project.html.

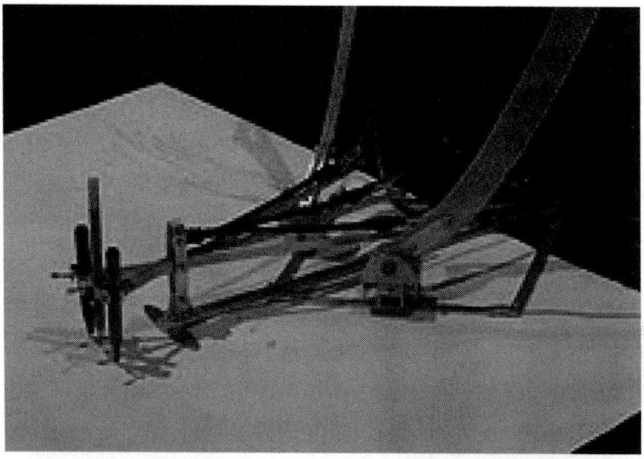

Fig. 12.10. Phil Gamblen, *MEART*, a single robotic arm that could draw.

12.5 Fractal Geometry: The Paradigm Shift in Art and Science

Recently we have seen a paradigm replacement both in art and science. Euclidean geometry has been replaced by fractal geometry of Benoit Mandelbrot in 1975 [Mandelbrot, 1977]. [7] Fractal theory offers methods for describing the inherent irregularity of natural objects. In fractal analysis, the Euclidean concept of "length" is viewed as a process. This process is characterized by a constant parameter known as the fractal dimension. The fractal dimension can be viewed as a relative measure of complexity, or as an index of the scale-dependency of a pattern.

Mandelbrot showed how fractals can occur in many different places in both mathematics and elsewhere in nature, rediscovering Leibniz's intuitions (1646-1716). [8] For instance, the fractal cube (Fig. 12.11) projects the grid plan of Renaissance perspective into the space of higher dimensions and complexity generated by contemporary mathematics. The fractal cube is an attempt to build the space of the new paradigm.

[7] Mandelbrot introduced the term "fractal" (from the latin *fractus*, meaning "broken") to characterize spatial or temporal phenomena that are continuous but not differentiable. For an elementary introduction to fractals, see [Lam, 1998].

[8] Leibniz intuitions: "matter presents an infinitely porous texture that is sponge like or cavernous but devoid of empty space, as there is always a cavern within every cavern."

Fractal objects are built like living trees, by successive repetitions. The challenge consists now of building a space with sufficient complexity to adequately reflect the complexity of the real. In the fractal cube, two geometrical paradigms—fractal space and Euclidean space—coexist simultaneously.

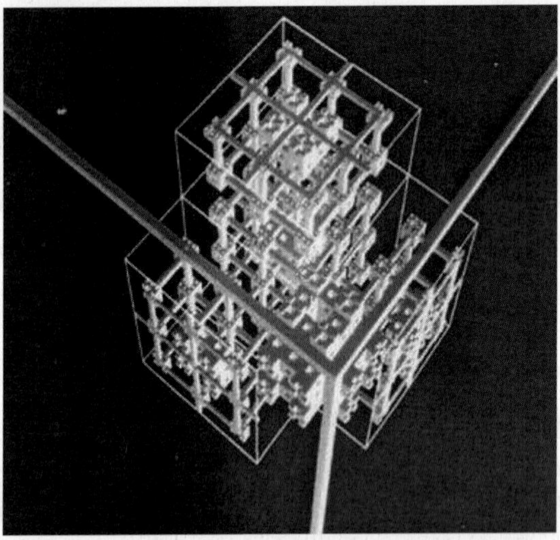

Fig. 12.11. *Fractal Cube* (1992), computer drawing, 9 x 9x 9 ft—a variation of the Sierpinski Cube.

Fractal geometry (e.g., Figs. 12.11 and 12.12), according to which every fragment of an object has the same or similar structure as the whole, can explain the structure of many of the objects that make up our universe: galaxies, the moon, the Earth, the atmosphere, the ocean, waves and clouds. Fractal geometry results from a complex work of architecture and sculpture based on several notions: repetition, identity, difference, direction, symmetry, open and close spaces, frontiers and angles, finite and infinite, the form and its absence [Salat & Labbé, 1994].

Curiously, there is an analogy between fractal representation and Vieira da Silva's (1908-1990) abstract paintings[9] as we can see in Fig.

[9] This artwork can be seen at Modern Art Centre in Calouste Gulbenkian Foundation and at Arpad Scenez and Vieira da Silva Museum, both in Lisbon, Portugal.

12.13. In her paintings there is also the idea of a space that is at the same time broken and continuous; this is a very contemporary idea, the idea of labyrinth.

Fig. 12.12. Serge Salat and Françoise Labbé, *Aleph II* (1989), mixed-media installation, 18 x 18 x 9 ft, Grand Palais, Paris.

Fig. 12.13. Vieira da Silva, *Léchec et Mat* (1949), 42.3 x 32.6 cm (left); *La Composition* (1948), 71.0 x 51.8 cm (right); both: oil on canvas,

12.6 Discussion and Conclusion

Here are some of our observations concerning art and science:

1. Creativity is a paradigm for both science and art.

2. The world is made up of polarities (good/evil, natural/artificial, static/dynamic) and we try to achieve our own balance. Art and Science are not exceptions.

3. The world must understand that scientists also react emotionally to their work—they also use their emotions to do science. However, the way they express it in scientific articles (instead of pictures as in arts) suppresses its emotional content.

4. Science is connected to the material world, while Art is more connected to spirituality, not in the religious sense but in the emotional sense. In Art, things are understood through emotions and not from the building blocks of formal explanations, as it happens with science.

5. The world of arts is much more related with aesthetic and beauty as well as the understanding of deep emotions.

6. The ultimate aim of science is to look for a single unifying "theory of everything"; Art does the same from an iconographic point of view, and includes spirituality.

7. The *noosphere* (i.e., sphere of human thought) today is growing exponentially in knowledge; it encourages as well as depends on the mobility of knowledge and the fact that we are all virtual editors of our collective discoveries.

8. Knowledge has in turn become its own object of inquiry: What constitutes knowledge? How should we use it? What kind of world will our new technology-driven Tower of Babel aspire to?

9. The sciences and technologies that enhance human physiology are becoming a vital part of artistic practices. Artistic practices utilize robotic computer-generated interfaces in biological arts for enhancing biological chemistry and bodily structures, and in BioArt for engineering and cloning non-human cells. Yet, these artistic works are not associated with the broader vision of human futures whereby the human is transformed beyond the biological condition. The role of human enhancement is to engage aesthetics and artistic participation in the evolving transformation from human to cyborg, trans-human, and later,

post-human. The human transformative sciences and technologies include nanotechnology, biotechnology, information technology, cognitive and neurological sciences for expanding cognition and extending physiology. Such enhancement does not mean purging human genealogy or discarding human values, nature, or the larger ecosystem of the Earth and all life forms. Instead, enhancement means supporting and assisting the human species and other life forms in overcoming disease, disability and injury, and for sustaining life. Emergent technologies of artificial general intelligence and molecular manufacturing and neurological sciences will widen possibilities for human enhancement, especially as these sciences and technologies converge.

10. BioArt seems to emerge at a time when the question "What is being?" is being posed in very innovative ways. The *MEART* project (The Semi Living Artist), currently being developed and hosted by SymbioticA, explores epistemological, ethical and aesthetical issues concerning the use of living neurons for ethno-centric end. Bio- and digital technologies are changing forever our concepts of the finite body by continually extending its capabilities.[10] Talking about BioArt and creativity, Christopher Auretta [2007] said:

> Digital implants, surgical transplants, biochemical modifications and our placement within an increasingly digitalized noosphere permit the finite and geographically determined body to become, in its online re(incarnation), infinitely ubiquitous and mobile.

How are we going to interact with such cybernetic entities considering the fact that their emergent behavior may be creative and unpredictable? How will society treat notions of artistry and creativity produced by semi-living entities? BioArt now mediates our increasing awareness that a dualistic philosophy operating within science deforms rather than informs the human. It is at the crossroads of such profound inquiry that BioArt seems to be shaping its realm of reflection and intervention. A bio-poetics should therefore be elaborated, we believe, in the light of

---

[10] In an online interview in 1999, David Cronenberg stated, "Modern technology is more than an interface. We ARE it. We have absorbed into our bodies."

these questions which BioArt consistently treats with epistemological delicacy, philosophical irony and ethical attentiveness.

11. Why is this taking place? The desire for new experiences is driving human enhancement forward. Coexisting in the virtual metaverse, interacting within gaming platforms, and building synthetic life forms have allowed us to experiment our existence in real time and synthetic time. As we continue to developed deeper communications and opportunities for experiencing the world around us, our creative explorations will undoubtedly grow further and reach farther.

References

Auretta, C. [2007] "Reflections on an emerging bio-poetics," in *Ciência e Bioarte—Encruzilhadas e Desafios Éticos* (Science and BioArt—Ethic Crossroads and Challenges)—*Caleidoscópio Edição e Artes Gráficas, S.A.* (Portugal) pp. 109-124.

Burguete, M. [2008] "History and philosophy of science: Towards a new epistemology," in *Science Matters: Humanities as Complex Systems*, eds. Burguete, M. & Lam, L. (World Scientific, Singapore) p.136-152.

Careri, G. [1994] "Discovering by perceiving in art and in science," *Leonardo* **27**(3), 181-182.

Charbonier, G. [1980] *Le Monologue du Peintre* (Neuilly sur Seine, Guy Durier Editeur, Paris).

Daudel, R. & Agaggio, N. [1994] *La Science et la Métamorphose des Arts* (Presses Universitaires de France, Paris).

Hargittai, I. & Hargittai, M [1994] "The use of artistic analogies in chemical research education," *Leonardo* **27**(3), 223-226.

Hoffmann, R. [2009] "Abstract science?" *American Scientist* **97**, 450-453.

Kemp, M. [2000] *Visualizations: The Nature Book of Art and Science* (Oxford University Press, Oxford).

Lam, L. [1998] *Nonlinear Physics for Beginners: Fractals, Chaos, Solitons, Pattern Formation, Cellular Automata and Complex Systems* (World Scientific, Singapore).

Mandelbrot, B. [1977] *The Fractal Geometry of Nature* (Freeman, New York).

Menezes, M. [2004] "Functional portraits: Picturing the invisible body," *A Minima* **7**, 6-17.

Prelog, V. [1976] "Chirality in Chemistry," *Science* **193**, 17-24.

Salat, S. & Labbé, F. [1994] "The Fractal Cube and the paradigm shift in art and science," *Leonardo* **27**(3), 241-248.

Snelson, K. [1994] "An artist's atom," *Leonardo* **27**(3), 231-236.

Soto, J. [1994] "The role of scientific concepts in art," *Leonardo* **27**(3), 227-230.

PART III

Understanding Arts

13

On the Origin of Literary Narrative and Its Relation to Adaptation

Patrick Colm Hogan

This chapter takes up the relation of literary narrative to adaptation. Due to the species-wide nature of most biological adaptations, this involves a particular focus on the universal features of narratives. The first section defines several senses in which we may say that a complex practice, such as literary narrative, is "adaptive." Subsequent sections have two goals. The first is negative. Literary narrative does not appear to be an adaptation in the strict sense, though there is some, very limited evidence that it may have "ancillary" adaptive value. The second, positive goal is to indicate how we may, in part, understand the origin of literary narrative—or, more precisely, the sources of many universal properties of literary narrative—as a complex result of the interaction of various evolved systems and processes. Both arguments suggest that the term "by-product" might best be reserved for a particular type of biological phenomenon such that literary narrative is more appropriately categorized as (in part) "adaptation-derived" rather than as a by-product. The final section considers what consequences this analysis has for assessing the value of literature and the arts. This arises as an issue since many advocates of the arts and humanities seem to view claims of adaptive advantage as crucial for establishing the worth of art. This section concludes with an argument that seeking the value of art in adaptation is misguided.

13.1 On Literature and Adaptation

In the last decade or so, there has been a good deal of debate over the origin of the arts, particularly the degree to which the arts may be spoken of in terms of evolutionary adaptation. [1] Indeed, in recent years, evolution—prominently, "Literary Darwinism"—has become the standard way of treating the origin of literary narrative. [2] But is literary narrative really the sort of thing that we can speak of as *evolved* in the technical meaning of the term? For example, we say that the human eye evolved and that the human visual system evolved. But can we reasonably say that literary narrative evolved in the same biological sense? Literary storytelling is not an organ, or a biological system, or even a set of specifiable behavioral propensities that derive from a system (in the way that flight/freeze/fight are behavioral propensities that derive from the fear system). Rather, literary storytelling is a complex and highly variable (if nonetheless patterned) *practice* that integrates a range of interacting contributory systems. It enhances, varies, and even suppresses particular system-based propensities. Can we speak of this sort of practice as having evolved?

13.1.1 *Principles of Evolutionary Explanation*

Before considering this, we should recall a few principles of evolutionary explanation that bear on the issue. The most fundamental principle is that evolution is the result of a genetic mutation that spreads throughout a population. The spread is typically the result of the functional consequences of the mutation, which is to say, the degree to which the mutation increases the selective fitness or adaptation of the organism in which it occurs. Adaptation here merely means the degree to which the organism is likely to pass on its genes to offspring. For example, a mutation that causes an increase in female fertility is adaptive since any females with the gene will be more likely to become pregnant. Other things being equal, that means that they are more likely to pass on their genes, including the fertility-enhancing mutated gene. The same point

[1] Perhaps the most widely known discussion of this issue is by Steven Pinker [1997]. For an influential response to Pinker, see [Carroll, 2004]. For a broader debate, see the special issue of *Style* (42.2/42.3 [2008]) on evolution and literary study.

[2] I use "literature" here in a medium-neutral way; thus it includes orature, film, video, etc.

holds for a mutation that enhances sexual allure, fosters behavior (e.g., attachment) that will help infants survive, and so forth.

Related to this, it is important to recognize the distinction between mechanisms and functions. *Mutations produce mechanisms*. In other words, they produce biological structures or processes that have particular causal consequences given certain conditions (e.g., certain environmental inputs). Functions are the general principles that explain the adaptive value of some mechanism. A mechanism has selective advantages if and only if it approximates some function with adequate frequency in the bodily, environmental, and social conditions in which it arises. But it is never identical with that function. This means that there are almost always conditions in which the mechanism does not conform to the function. For example, it is clear that the function of our fear system is to produce avoidance of danger. However, it is also clear that the fear system sometimes produces freezing when it should produce flight, sometimes produces freezing or flight when there is no danger at all, and sometimes produces no avoidance response when there is real danger. This is because the fear system involves a complex set of mechanisms that approximate the function; it does not include the function itself (avoiding danger).

The distinction is crucial for evolutionary thought. Yet it is often formulated in a misleading manner, sometimes with unfortunate consequences. For example, in *The Structure of Evolutionary Theory*, Stephen Jay Gould repeatedly characterizes the crucial explanatory distinction as that between efficient causality (causality in our ordinary, scientific sense) and final causality (i.e., causality formulated in terms of actions guided by goals) [2002, p. 1187]. Function perhaps makes more immediate intuitive sense when formulated in terms of goals. Thus Gould writes that "feathers first evolved for thermoregulation" [2002, p. 86]. Phrasing this development in terms of a purpose makes it readily comprehensible. But, of course, there is no goal at all, no final causality. Rather, there is only an analogical relation between goal-oriented behavior, on the one hand, and the operation of natural selection, on the other. Mutations led to proto-feathers, which had a selective advantage due to thermoregulation. We use the analogy—feathers developing as if on purpose—in order to make intuitive sense of the purely efficient causal processes operating through mechanisms.

The difficulty here is that failure to maintain a strict mechanism/function distinction easily leads to false inferences. Thus Gould writes that, "As functionalist theories, both Lamarckian soft inheritance and Darwinian natural selection share a defining premise that environmental information about adaptive design somehow passes to organisms, and that organisms then respond by fashioning traits to enhance their competitive ability" [2002, p. 1179]. This is very misleading, if not simply false. Organisms do not respond to the environment by fashioning traits. If anything, they "fashion" traits (through mutation), then "respond" to the environment (by successfully passing on their genes or not)—though that too is merely an analogical way of putting it.

Even Darwin himself sometimes treated functional approximation as if it were actual final causality. Thus we find him writing that "As natural selection works solely by and for the good of each being, all corporeal and mental endowments will tend to progress towards perfection" [1872, p. 489]. If we think of evolution in terms of goals, this claim of a movement toward perfection may seem to make some sense. However, if we recognize that evolution is a matter of mechanisms operating to spread genetic mutations in a particular environment, then we recognize this as a *non sequitur*.

Conceptual obscurity and inferential errors bear not only on adaptations, but also on *by-products*. It is a commonplace that every feature of human genetic development is either an adaptation or a by-product. This would entail that literary narrative is either an adaptation or a by-product. In my view, this division is theoretically problematic. Primarily, it is overly broad. For example, it would seem to entail that the same behavioral propensities (e.g., freezing in the face of sudden approach) are adaptations in one context (where they approximate a function), but a mere by-product in other contexts (where they do not approximate a function). I would, instead, reserve the use of the term "by-product" for features that are operationally independent of the adaptation. For example, suppose a certain genetic mutation has consequences in language-related areas of the brain but also in the lungs (as occurs with *FOXP2* [Marcus & Fisher, in press]). If the former produce mechanisms with adaptive functions, but the latter do not, then we may say that the changes in the brain are adaptations, but the changes in the lungs are a by-product. I would distinguish by-products in this sense from both adaptive mechanisms and adaptive functions.

Finally, it is often the case that some complex practice results from the interaction of the usual mechanisms of adapted systems. Though the practice goes beyond any single component system, the operations or mechanisms are the same as those of the components. For example, standing on one's hands on the back of a trotting horse uses adapted vestibular, postural, and locomotion systems, along with adapted cerebellar and other processes for the acquisition of motor routines, the adjustment of those routines in particular circumstances, and so on. It clearly goes beyond these systems considered singly. It even goes beyond such adapted cross-system integrations as occur in the development of bipedal locomotion. But it is still nothing more than a particular combination of those adapted components. (The same point presumably holds for a wide range of practices in human life—perhaps virtually all of culture.) Thus practices of this sort are not by-products in my sense. But they are also not adaptive mechanisms per se; they are, rather, the result of adaptive mechanisms. I will refer to such complex practices as "*adaptation-derived*."

13.1.2 *Evolution and Complex Practices*

This leads us to the issue of when a complex practice is itself evolved or adaptive, thus not simply adaptation-derived. An obvious criterion for an adaptation would be something along the following lines: A particular practice is adaptive if and only if it is possible to isolate genetic mutations that were necessary to the development of some mechanism of the practice and that mechanism approximated a function for reproducing the mutated gene in the bodily, environmental, inter-personal, and social context of the person with the mutation. That fits well enough with our understanding of evolution. Unfortunately, it will not do as an account of adaptation for the simple reason that it is too broad to be useful. It seems very unlikely that there is anything about humans that does not involve some adaptive mechanism. For example, this definition would count, not only literature, but everything from political assassination to "pull my finger" jokes as adaptations simply because they involve planning. Since the ability to plan is presumably evolved, anything that makes use of that ability becomes adaptive by this criterion. Clearly, we want an account that is more fine-grained. (I should note that the point here is far from trivial. It is not uncommon for humanists to think they have made a case for the evolution of literature or the arts when they have only made a case for the evolution of some component of literary production or reception.)

1. Strict adaptation and ancillary adaptation

So, how do we solve this problem? The difficulty with the preceding criterion is that the adaptive advantage manifest in the derived activity (assassination, "pull my finger" jokes, whatever) need not go beyond the adaptive advantage of the component activity (in this case, the capacity to make plans). There are two ways of responding to this.

First, we may require that there be some genetically determined mechanism that spreads due to selective advantage and that it does so only in connection with the target activity. In our case, this would be a mechanism that spreads solely in connection with literary narrative. In other words, this mechanism is not functional in any of the component systems (e.g., working memory). I will refer to this as "adaptation in the strict sense" or "*strict adaptation.*"[3]

[3] I should note that my usage here differs from that of Gould. A "strict adaptation" in my sense may be either an "adaptation" or an "exaptation" in Gould's sense. For my purposes, it does not matter whether the initial appearance of a mechanism occurs separately from the activity in question. The crucial issue is whether it has already spread as far as it will spread in the population or whether the activity (e.g., narrative) leads to the trait spreading due to selective advantage. For example, suppose a subset of people had a particular genetically determined feature before the development of narrative. Initially, this feature produced no selective advantage. Then, after the development of narrative, it began to produce a selective advantage and spread throughout the population. This would count as a "strict adaptation" in my sense even though the trait existed before the activity (narrative) made it approximate a function and gave it a selective advantage.

Less restrictively, we may require that the incorporation of a prior adaptive mechanism into the target activity (e.g., literature) serve to distinctively enhance the reproductive benefits of that mechanism. In this case, the selective fitness of anyone engaging in the activity would be enhanced. However, the prior adaptation may already be universal in the population, in which case the new practice would not select for that particular trait. For example, suppose literature distinctively enhances the adaptive value of working memory. Then people who read literature may be more likely to pass on their genes (if, say, enhanced working memory gives them advantages in acquiring resources). However, since working memory is already universal within humanity, that selective advantage would not further the presence of any particular gene contributory to working memory or any biological trait resulting from that gene. I will therefore refer to this selective advantage as "*ancillary adaptation*."[4]

Note that I have explained ancillary adaptation in terms of a "distinctive" contribution to selective fitness. I say "distinctive" because a given trait might in principle be improved by practice and any form of practice may be as good as any other. Suppose thinking about plays in football, geometrical proofs, themes in literature, gossip, theology, and many other things all allow us to improve working memory operation. Then it does not seem to make sense to say that all these activities are adaptations, even ancillary adaptations. In contrast, we would be likely to give special status to a particular sort of activity if it proved to be distinctively—in the best case, uniquely—suited to enhancing the selective advantage of the prior mechanism. If literary study contributes to working memory only as much as studying football or paying close attention to gossip, then we would probably be disinclined to say that it has adaptive value, even ancillary adaptive value. However, if it is unusually well suited to the task, then it makes sense to credit literary study in this way.

---

[4] This need not involve a shift in function of the sort stressed by Gould. I suspect that it is more likely to involve an intensification of the already existing functional approximation.

2. Literature, strict adaptation and ancillary adaptation

What consequences does the first, more restrictive requirement have in the case of literature? If literature is strictly adaptive, then the genetically governed features of literary narrative cannot be explained fully by mechanisms deriving from component systems (such as those involved in planning, memory, or emotion). Rather, there must be at least one literature-specific genetic feature that produces a selective advantage. Conversely, literature is not strictly adaptive if all its genetically-governed features may be accounted for by component mechanisms (i.e., mechanisms of prior systems) or if any remaining features do not yield a selective advantage. Outside of literature, language may be an example of a complex practice (or capacity/system plus practice) that is adaptive in this sense. Clearly, language draws on working memory, long-term procedural memory, perception, and other adapted systems. But writers such as Chomsky have argued that there is a distinctive genetic component to language as well. If that distinctive component has selective advantages (as would seem likely), then language counts as strictly adaptive.

What about ancillary adaptation? In that case, literature would not add any adaptive mechanisms. However, it would distinctively enhance the selective advantage of one or more pre-existing mechanisms—such as working memory, in the hypothetical case mentioned above. An example of an "ancillary adaptive" practice might be yoga. It seems unlikely that yoga involves any unique adaptive mechanisms. However, it may be that yoga enhances the adaptive operation of such pre-yogic mechanisms as the modulation of emotion activation by respiration and posture (as opposed to the modulation of respiration and posture by emotion activation). It may be that there are few other systematic practices that take up this adaptive "reverse causality" of respiration and posture, enhancing their benefits. In this case, yoga would be distinctive.[5]

[5] The most obvious adaptive value of respiratory and postural "feedback" to emotion system activation is in connecting our spontaneous mirroring of other people's posture and respiration with emotional response and thus enabling emotional contagion (see [Hatfield et al, 1994]).

In the following sections, I will consider the origin of literary narrative and the relation of this to strict, then ancillary adaptation.

13.2 Literary Narrative and Strict Adaptation: What Makes a Story

My argument in this section is, first of all, that many universal features of literary narrative may be explained by adaptive mechanisms.[6] Thus the systems that manifest these mechanisms help to explain the origin of—or, more properly, provide sources for—literary narrative. However, that does not make literary narrative adaptive. The specifically genetic origin of literary narrative appears to be entirely accounted for by the operation and interaction of the prior mechanisms. There does not appear to be a further adaptive mechanism that is distinctive of literary narrative. At the same time, by the preceding definition of "by-product," literary narrative is not a by-product, but "adaptation-derived," sharing the same (function-approximating) mechanisms as its sources.

13.2.1 *Literary Universals*

A key issue to consider here is just what features of literary narrative are candidates for genetic determination. These would be, first of all, any narrative universals, thus practices that are apparently not the product of cultural or historical variability. I cannot go into the evidence in a short paper. However, I have argued at length in *The Mind and Its Stories* [2003] and *Affective Narratology* [in press] that there are important narrative patterns that recur across unrelated traditions. These properties

[6] I say *many*, not all. Since this chapter focuses on evolutionary accounts, I should stress that universal patterns may arise from various sources, including aspects of the environment; principles of complex systems (see Lam [2008] for a case of this sort from history); features of group dynamics; occurrences in childhood development that are cross-culturally recurrent, but not genetically determined, etc. In other words, evolutionary explanation does not bear on all universal patterns, but only on those universals that have a genetic basis (thus not a basis in physics, group dynamics, etc.). This is widely known. Yet it also appears to be widely forgotten. Particularly in the study of literature, it seems to be often assumed that any cross-cultural aspect of narrative must be biologically determined.

include the following: First, stories are told for the purpose of emotional engagement in perhaps every society.

Second, these stories may or may not have sources in fact, but the telling of the focal events is often clearly dissociated from any direct, factual concern (i.e., it does not serve primarily as a literal source of information on such events). (Generally, this means that they are or are treated as fictional.)

Third, the general format of a story involves beginning with some sort of context-relative normalcy. That normalcy is disrupted in a partially aversive way. The new condition generates the formation of a goal on the part of one or more protagonists. This is followed by goal pursuit in which an agent encounters events and circumstances—prominently intentional acts by other agents—that further or inhibit his or her goal pursuit. At some point, the narrative commonly includes the apparently complete and irreversible loss of the goal. In tragic stories (which are much less common than comic stories), this concludes the sequence, establishing a new normalcy. In comic stories, the goal is ultimately achieved, also establishing a new normalcy. This comic normalcy is typically idealized in relation to the goal (e.g., if the goal is marriage, then the couple lives "happily ever after").

Fourth, many stories follow a more specific, genre pattern as well. Cross-culturally, by far the most common genres are heroic, romantic, and sacrificial tragi-comedies. Stories concerning parent/child separation and reunion, sexual pursuit, and revenge or social justice also occur across cultures, but less frequently and less prominently. All these genres have clear, recurring patterns.

13.2.2 *Imagination and Literature*

The basic questions of the origin—or sources—of literary narrative arise here: To what extent can these properties be explained by reference to routine cognitive and affective architecture (leaving aside group dynamical and other non-genetic causes for the present inquiry) and to what extent (if any) do they include distinctive mechanisms?

Perhaps the most important component of literary narrative production is what literary critics and theorists refer to as "imagination" and cognitive scientists and evolutionary psychologists refer to by such labels as "off-line processing." Many writers on literature and adaptation (e.g., Tooby and Cosmides [2001]) have remarked that the arts involve off-line processing—the imagination of hypothetical or counterfactual conditions and events—and that off-line processing is adaptive. Both points are certainly true. As to adaptation, imagination allows me to envision possible routes to success in, say, hunting. I may then follow through one of these routes in reality. It also allows me to envision possible routes to failure so as to avoid these (cf. [Wong, 2000, p. 10]). The general relation between off-line processing and literary narrative is fairly self-evident. But the connections go beyond what is commonly recognized. They are not simply a matter of both being hypothetical. The precise nature of the imagination is closely parallel in the two cases. In the main adaptively significant cases (such as imagining successful or failed hunting), off-line imagination comprises precisely a causal sequence in which an agent pursues a goal through varying the conditions and events—prominently including acts by other agents—that would help or hinder the achievement of that goal. In this way, the trajectory of practical off-line imagination is virtually identical with the basic trajectory of prototypical literary narrative.

Of course, literary imagination is not usually autobiographical in a literal sense (i.e., the protagonist is not usually the author per se). But that is not necessarily different from practical or "basic" off-line hypothetical imagination. In ordinary life, off-line imagination operates not only egocentrically but empathically as well. We imagine that other people have goals and we imagine that they will work to achieve them—particularly insofar as this work will help or hinder our own goal pursuit. Indeed, this is crucial to the adaptive value of imagination. Perhaps more significantly, even our imagination of ourselves is, in effect, empathic as when we imagine different ways in which we might react to a hypothetical situation or different properties we might acquire (e.g., how I would handle a particular situation if my grasp of Hindi were firmer).

Obviously, this brief summary of our off-line imaginative processing of hypothetical actions leaves a great deal unexplained. But, for our purposes, the crucial thing is not how we account for off-line imagination. Rather, the crucial thing is that off-line imagination explains a great deal about literary narrative. It seems clear that it accounts for one significant part of the origin of literary art. Does it then indicate that literature is adaptive? Even the barest connection between off-line processing and literary imagination seems to strike many people as evidence that literature is adaptive. However, the detailed connections between the two actually indicate the opposite. By our strict standard, the imaginative development of counterfactual goal-pursuit is adaptive in its general, non-literary operation. Thus its operation in literary narrative cannot be counted as indicating that literature is strictly adaptive. The adaptation precedes the literature and literature is, at least in this respect, the result of the (adapted) cognitive architecture. It is, then, adaptation-derived.

13.2.3 *Literature and Emotion*

But there is an obvious, or at least apparent, difference between off-line processing and literary narrative. They have different purposes. Specifically, how are we to account for the enjoyment of hypothetical imagination in literature? Our ordinary off-line processing explains much of the basic structure of literary stories. But what about the pleasure?

In fact, this too appears to be part of the same prior adaptation. Specifically, it seems clear that some motivational system—perhaps some aspect of the *endogenous reward system*—is involved in ordinary, practical acts of imagining the causal elaboration of goal pursuit (i.e., ordinary, practical off-line processing). I say this because a great deal of the selective advantage of off-line imagination derives from imagining aversive outcomes—as when I imagine possible dangers that I may then avoid. Even the imagination of positive outcomes is adaptive only insofar as it takes into account possible failures. But aversive outcomes are, precisely, aversive. As such, we should not only shun them in the real world, we should want to keep from thinking about them. Why, then, do we imagine them?

Suppose, for example, that Glug is thirsty. He imagines going to the river for a drink, then follows through on this as an imagined plan. This is adaptive only insofar as Glug actually allows for the possibility of, for example, predators being at the river or at particular points on the river at particular times (which he may then avoid). In other words, it is adaptive only if Glug includes something unappealing in his imagination. Indeed, if Glug does not include such aversive aspects, we are likely to refer to his hypothetical thought as "fantasy" rather than as "imagination."

13.2.4 *Explaining Aversive Imagination*

One question we need to answer, then, is—why would we ever engage in (potentially aversive) imagination rather than (enjoyable) fantasy? Three reasons come to mind. First, we may have aversive experiences resulting from mere fantasy. I engage in fantasy, act on the fantasy, and, as a result, have a bad experience (e.g., barely escaping from the predators at the riverbank). Such bad experiences are stored as emotional memories. (Emotional memories are memories that, when activated, reproduce the initial emotion—for example, fear.) Subsequently, those memories discourage me from following through on plans based merely on fantasy. But this only explains why we would avoid action based solely on fantasy. It does not explain why we would actually imagine aversive conditions (e.g., being attacked by a predator).

The second reason to engage in the full emotional spectrum of imagination is that we take pleasure in problem solving. This may be, for example, some sort of anticipatory dopaminergic operation. In any case, pleasure in problem solving goes some way toward explaining our off-line imagination of aversive situations. It is also one clear source for our enjoyment of literary narratives. But I do not believe it completely explains either. As to literary narrative, there are obvious aspects of our enjoyment that are independent of problem solving. Otherwise, it would be difficult to re-read the same story with pleasure many times, long after we have figured out the narrative puzzles.

As to off-line imagination outside literature, not all instances involve problem solving. For example, part of the adaptive function of imagination is "being prepared" for possible threats. There is no real problem-solving enjoyment in these cases because we already know how we should react. For example, before meeting a student to discuss a grade, I may imagine that he or she could be hostile about the grade. I imagine the situation beforehand in order to keep calm if this occurs. There is no problem-solving pleasure here because I do not have any uncertainty to puzzle out. Why, then, do I engage in the imagination in such a case—since, again, it is aversive? It does no good to say that it is in my best interest, that it is practically beneficial, since that only names the function of the imagination, which we already know. The issue is what mechanism underlies this function.

I would suggest a third reason—the mechanism involves a distinct enjoyment that results from the detailed imagination of emotionally consequential causal sequences involving agency. Clearly, this is a notion that needs to be developed in algorithmic detail. It is in part connected with attentional orientation, thus the motivational rewards of "interest," but it does not seem reducible to these. In any case, it seems clear that something of this sort must underlie and motivate off-line imagination in real life. Put simply, there is an adaptive advantage to experiencing some pleasure at the imagination of even aversive outcomes. The point fits well with our enjoyment of the detailed imagination that is central to literary narrative.[7] (It may also be related to other phenomena, such as the fascination that draws people to gawk at disasters, such as car accidents.)

13.2.5 *Explaining Literary Emotion*

Here, too, we explain something about the sources of literary narrative. Indeed, by incorporating a motivational component, this account indicates that the development of something along the lines of literary narrative was virtually inevitable. At the same time, this account leads us to expect certain sorts of complexity in literary narrative. In other words, it has consequences beyond the initial explanation.

Specifically, given the preceding analysis, we would expect people to experience a conflict between the straightforward pleasures of fantasy

[7] That detailed imagination has been explored particularly by Elaine Scarry [1999].

and the sterner enjoyments of imagination. In cases where positive outcomes are likely, the conflict is limited. Moreover, in literary narrative, there is no direct practical consequence (such as being eaten by a predator) that may result from opting for fantasy. Thus the problems with fantasy (including the painful emotional memories) do not arise. As a result, we might expect comedies to be predominant in literature, and we might expect fantasy to play an important role in literary narrative—which is just what we find (e.g., in the idealization of comic outcomes).

On the other hand, we might also expect the fantasy/imagination conflict to recur in literature—and, indeed, we appear to see just that in the common opposition between escapism, on the one hand, and "serious" fiction (including certain forms of realism) on the other. Moreover, we would expect some literature to stress aversive imaginations, since these do occur spontaneously in ordinary life and must have their own (adaptive) motivations. This is, of course, what we find in tragedies.

Thus we see that a number of fundamental, universal features of literary narrative appear to have their source in the structure and emotional complexity of off-line, hypothetical imagination. In other words, we are able to give a fairly straightforward account of at least some aspects of the origin of literary narrative. In these respects, literary narrative appears to be adaptation-derived, thus neither a strict adaptation nor a by-product. Of course, off-line imagination does not account for every universal feature mentioned above. For reasons of space, I cannot treat all of these. I have, however, discussed the remaining features in [Hogan, 2003; in press]. For example, in these two books, I argue that the nature of the beginning and of the usually comic ending found in standard story structures, as well as the usually tragic middle, result directly from the nature of our emotion systems. The same point holds for the recurring structures of the cross-cultural genres and the relatively greater frequency of the heroic, romantic, and sacrificial structures. In these cases, too, then, it appears that we may explain many universal features of stories—thus, at least in part, the origin of literary narrative. To a great extent, we may do this by reference to the complex interaction of prior, adapted systems. For a complete account, we would need to add group dynamical and other factors. However, there does not seem to be any reason to posit an adaptation arising with literature itself.

13.2.6 *Literary Narrative Beyond Imagination: Explaining Discourse*

Before turning to ancillary adaptation, we should briefly consider one further area in the study of literature. As narratologists would insist, stories are not all there is to narrative. The other area of narrative is "discourse," the manner in which stories are presented. Here, we might consider two key components of discourse—the recounting of the events in the story and the perspective from which they are recounted. Here, too, the question arises as to how we might explain cross-cultural patterns in these areas.

As to the events recounted, stories may be told in chronological order or outside chronological order. Most often, they are recounted in largely or entirely chronological order.[8] As to the perspective of the telling, the events may be recounted by different sorts of narrator. The usual varieties are an omniscient narrator outside the story world or a limited narrator who is part of the story world.

In fact, given our preceding treatment of literary narrative and imagination, this is just what we would expect. Our practical off-line imagination of goal-pursuit typically unfolds in chronological order. Moreover, it involves both an omniscient perspective outside the story and a limited perspective within the story. Specifically, as the consciousness that is imagining possible trajectories, I am in control of determining just what the events are. In that sense, I am "omniscient" with respect to the imagined world.[9] But, at the same time, my imagination involves a limited, story-internal perspective as well—that of me as the object of my imagination. For example, I might imagine trying to cross the road after my glasses have been broken. From the

---

[8] As we would expect from the preceding analysis, this is true within the trajectory of action of a single character. There are complications when more than one such trajectory enters.

[9] Of course, I am not omniscient with respect to the real world. But that is precisely the aspect of practical imagination that does not enter with fiction.

omniscient perspective (me as imagining consciousness), I imagine a car coming. Simultaneously, from the limited perspective (me as imagined goal-pursuer within the imagination), I do not seeing the car. Much of the adaptive value of off-line imagination actually derives from this dual perspective. These properties of hypothetical imagination serve as the sources of the parallel properties in literary narrative.

Note, however, that chronological order and the use of these particular types of perspective, though standard in both practical off-line imagination and literary narrative, are not necessary in either case. As to practical imagination, it is clear that we synthesize fragmentary information all the time. For example, in hearing reports about other people's activities, we often have to (imaginatively) piece together information concerning different points in a jumbled temporal sequence. Moreover, that information derives from tellers with different sorts of knowledge, with different relations to the events of the story, and different degrees of reliability. Thus there is always the possibility of non-chronological narrative and there is always the possibility of other sorts of narrators, some of whom are inconsistent, some of whom are limited in knowledge, but outside the story, some of whom are unreliable, and so on.

Thus in literary discourse too we are able to isolate sources for cross-cultural patterns, thereby explaining these aspects of the origin of literary narrative. These sources are to be found in prior propensities, systems, etc. The prior propensities (e.g., imagining causal sequences chronologically) almost certainly have selective advantages. However, here too, there does not appear to be anything distinctive of literary narrative that would make it an adaptation in the strict sense.

13.3 Literary Narrative and Ancillary Adaptation

We have been speaking about the sources of literary narrative. But what happens once one has literature? Is it possible that literary narratives take up adaptations and render them more effective in a distinctive way? In other words, does it happen that literary narrative actually produces better versions of the prior adaptations, versions that would not be available (or would be available only with great difficulty) outside literature?

Consider a possible case. It is well known that our ordinary theory of mind capacities allow only narrowly limited embedding. Thus, we can understand "Jones thinks that Smith believes that Doe loves Mary." But if we add only a couple of minds, we get lost—"Williams expects that Robertson hopes that Jones imagines that Smith believes that Doe loves Mary." A five-level embedding of this sort is, it seems, beyond our capacities. However, Lisa Zunshine maintains that, in literary reading, our theory of mind capabilities are extended even to six-level embeddings [2006]. If this is so, could literary study increase the adaptive value of theory of mind?

In fact, Zunshine explicitly rejects the idea—"reading of fiction does not make us into better mind-readers," she writes [2006]. But, if literature does lead us beyond our usual degree of embedding, it seems possible—even likely—that it could enhance the adaptive value of theory of mind. Of course, it would not change the actual adaptation. Suppose I read literature and increase my competence at embedding. That may make me more fit for reproduction. For example, I may be able to manipulate groups better, thereby succeeding more in my career, thereby gaining greater resources for my own health and that of my children, etc. But this would not produce any genetic change in me or in my offspring. It would not alter my genes, thus producing a new adaptation. Rather, it would enhance the adaptive consequences of a prior adaptation. Is that the case here?

Zunshine has done literary criticism and theory a great service by bringing theory of mind research into the analysis of character. However, her argument about increased embedding in literature does not seem to hold. Consider an example from Virginia Woolf's *Mrs. Dalloway*. Zunshine treats the following passage [(1925) 1981, p. 110]:

> Hugh produced his fountain pen . . . which had done twenty years' service, he said, unscrewing the cap. It was still in perfect order; he had shown it to the makers; there was no reason, they said, why it should ever wear out; which was somehow to Hugh's credit, and to the credit of the sentiments which his pen expressed (so Richard

> Dalloway felt) as Hugh began carefully writing capital letters with rings round them in the margin, and thus marvelously reduced Lady Bruton's tangles to sense, to grammar such as the editor of the *Times*, Lady Bruton felt, watching the marvelous transformation, must respect.

Zunshine sees this as involving as six-level embedding—thus a degree of embedding beyond the indecipherable sentence "Williams expects that Robertson hopes," and so on. As she explains, "Woolf *intends us to recognize* . . . that Richard *is aware* that Hugh *wants* Lady Bruton and Richard to *think* that because the makers of the pen *believe* that it will never wear out, the editor of the *Times* will *respect* and publish the ideas recorded by this pen" [2006, p. 33, italics in the original].

But, in fact, the passage communicates something different. First, even in Zunshine's own phrasing, the editor's respect is not embedded within the makers' belief. Thus her own statement only yields five embeddings at most. Second, the passage does not involve any positing of Virginia Woolf as relevant to embedding. Yes, Woolf is the author. Yes, we are concerned with Woolf's intentions in the passage. But that is no different from any other utterance—for example, me saying "John likes Mary." The utterer only becomes part of the embedding if his or her intentions are considered explicitly. There is no reason to believe that is happening in this passage. Given this, the embedding in the passage itself cannot be more than four levels deep. Moreover, even this does not seem to be born out by the text. Specifically, the passage in no way indicates that Hugh wants anyone to draw any conclusions about the preferences of the editor of the *Times*, which removes one further layer. Indeed, the entire passage does not represent a single set of embeddings in any way. Rather, we have one set of embeddings in which Richard *feels* that Hugh *believes* that the quality of the pen shows the value of Hugh's writings (two embeddings—the pen showing value does not involve any theory of mind). We might say that Richard *feels* that Hugh *believes* that the testimony of the makers indicates the value of Hugh's writings. If we take "testimony" to be a mental attitude, then we can squeeze three levels of theory of mind embedding out of this. But that seems to be the maximum. Then we have Lady Bruton *feeling* that Hugh's writing will be *respected* by the editor—a mere two levels of embedding. The passage is stylistically brilliant and psychologically complex. It is also difficult, due to some aspects of syntax and the subtle ways in which mental attitudes are attributed. But it does not involve any unusual degree of theory of mind embedding.

Moreover, even if we were to find cases of unusual depth of embedding in *Mrs. Dalloway* or elsewhere, that would not necessarily show that our theory of mind capacities are enhanced. If our capacities were enhanced by literature, then—in contrast with ordinary people—literary critics should be able to follow six-level embeddings wherever they occur (as in "Williams expects that Robertson hopes that Jones imagines that Smith believes that Doe loves Mary"). But that does not appear to be the case. (This may be one of the reasons why Zunshine rejects the idea that enhanced embedding is adaptively significant.)

However, not all cases of possible enhancements are like this. There is some evidence that literary study does improve the functionality of one set of prior adaptations—those clustered together under the rubric of "emotional intelligence" (e.g., identifying other people's emotions). Research by Keith Oatley and others indicates that reading literature has both short-term and long-term effects in this area [Oatley, 2009]. These are important results and potentially very significant. But, even here, it is not entirely clear whether or not this is merely an "exercise" effect. In other words, it may be that, by engaging in, say, emotion-inference, we become better at it. Literature would be one way of engaging in such "exercise" of our capacities. But the study of psychology cases or personal experiences might produce the same results. Moreover, it is not clear that even these effects entail any selective advantage. The study of literature could improve subjects' emotional intelligence without thereby increasing their reproductive success. Nonetheless, the results do suggest a potential role for literature in ancillary adaptation.

13.4 Productivity and Eudaimonia: On the Obsession with Literary Adaptation

In sum, the possibilities for explaining the origins of literary narrative seem very good. Indeed, a large part of that explanation, as outlined above, seems fairly straightforward. But the case for literary narrative as an adaptation does not seem promising. Literary narrative is, to a great extent, adaptation-derived. Thus it is not appropriate to refer to it as a by-product. But it seems highly unlikely that it is a strict adaptation. It may be an ancillary adaptation, but the evidence for this is not strong.

I imagine the response of most literary critics who read this argument will be one of either sorrow or outrage. As a profession, we seem to be deeply invested in the idea that literature is adaptive, even if we do not agree with the Literary Darwinists as to just how that might be the case. Yet adaptation is neutral on all value scales except that of reproduction. Being able to kill more efficiently is no less adaptive than being able to cooperate more effectively (depending on precise circumstances). I suspect that most literary critics would vote for cooperating more effectively rather than killing more effectively, even when the latter is more adaptive in the sense of increasing our reproductive capacities. Given this, why are we so obsessed with literature as adaptation?

A recent article in *The Nation* magazine suggests part of the reason why. Though generally critical of the Literary Darwinists, the author imagines their claims being reported under the headline "Researchers Find Arts Have Value" [Deresiewicz, 2009, p. 28]. Basically, the assertion that literature is adaptive gives teachers of literature and the arts a response to legislators who insist that literature is not useful. Indeed, William Deresiewicz conjectures that Literary Darwinism ideas may "restore the prestige of the arts and humanities, and . . . maybe even their funding" [2009, p. 28].

To understand and respond to funding difficulties, we need to understand just what counts as value for the legislators, and even for much of the general public. The dilemma facing defenders of literary study is that value is widely if tacitly conceived of as a particular sort of productivity. Basically, the value of any form of study, or of almost anything else, is reduced to productivity in the service of social hierarchies. This is not to say that it is productivity in the service of current elites (though there is certainly a significant bias in that direction). Rather, it is to say that a course of study is seen as having value if it gives the learner some hierarchical advantage over others—prominently, in wealth or institutional authority.[10] Darwinists today tend to be somewhat obsessed with the idea that, at least for males, social rank increases sexual desirability, thus the likelihood of having children. Thus anything that increases one's rank may be seen as adaptive. In this way, claiming that literature is adaptive is related to claiming that it fosters status-enhancing productivity. As a result, the claims of the Literary Darwinists may be seen as fitting the legislative needs of literature teachers.

But there are several problems with this view. First, once again, literature does not seem to be adaptive, but only adaptation-derived. It is not even clear that it has an ancillary adaptive function. I certainly believe that Oatley's valuable findings should be made available to administrators and legislators. But I do not believe that we should make unsupported claims about art and adaptation even if they would help secure funding for literary study.[11] Second, I am not at all sure that a circuitous link between adaptation and status productivity would help with PR (public relations) for literary study anyway. (Imagine the advertising slogans—"Read Milton! You'll have way more babies!" or "*Proust*—it's French for fertility!") Finally, and most importantly, I personally am loath to accept the equation of value with status productivity and I suspect most other literature professors are as well.

It is, however, possible to suggest an alternative to status productivity as a standard for the evaluation of literary narrative—Aristotle's *eudaimonia*, "human flourishing" (as Martha Nussbaum puts it [2001, p. 32]) or, more simply, living a good life. Suppose our goal in education is not preparing students to dominate more people more effectively, but rather preparing them to live well. Of course, living well involves having employment and income. Thus we cannot ignore job-

---

[10] A striking case of this may be found in Cacioppo and Patrick's book on loneliness. They explain the value of freeing oneself of loneliness by pointing out that doing so will "improve job performance, increase the likelihood of receiving good performance reviews and promotions, and provide better networking opportunities for career growth" [2008, p. 220]. Of course, some people might be lonely because they find something wrong with a social structure based on "performance reviews" that evaluate one's profit-enhancing productivity, in-group preference through networking, etc.

[11] I should stress that Oatley is not in any way making such claims.

placement for our students. But a good life is not confined to job-placement. Rather, it centrally involves such components as an appreciation of beauty and the ability to reflect on human problems and dilemmas, perhaps including the imagination of times, places, and people that go beyond one's ordinary experience. Indeed, eudaimonia may involve enhanced emotional intelligence in ways that do *not* increase reproductive fitness (e.g., in leading to forms of empathy-based altruism that are not reciprocal and not confined to kin). Our ability to experience beauty and to imagine the experiences of other people are evolved capacities. Yet their development through literature may not enhance the likelihood that we will pass on our DNA. At the same time, however, that development may make our lives better.

In an even more utopian vein, insofar as we value human flourishing over status productivity, we may be able to contribute toward a change in common practices. Our biological propensities are not absolute and invariable laws. Just as we can modulate our emotion systems, inhibiting anger or fear, we may modulate other tendencies as well. This may actually change what maintains adaptive value. For example, in the context of universal belligerence, being pacifistic may endanger a society and being belligerent may contribute to the survival of a society, thus the reproduction of its members. However, the survivability of all societies—and thus the reproduction of members of those societies generally—may be enhanced by pacifistic cooperation. Orienting our cultural practices toward some sort of universal eudaimonistic cooperation is perhaps more reminiscent of Kropotkin's vision [(1902) 1989] than that of most socio-biologists and Literary Darwinists.[12] That vision may be more in keeping with the interests and propensities of

[12] Kropotkin argued that a crucial contributing factor in evolution is "mutual aid" and linked this with his advocacy of egalitarian cooperativism. Evolutionary Psychologists certainly recognize the importance of cooperation. But they understand cooperation itself in a somewhat narrow and competitive way. Consider, for example, the view of Tooby and Cosmides that communal redistribution of wealth is selected against by evolution (1992: 213). Social cooperation is necessary for preventing such redistribution. This is clearly not the sort of cooperativism that Kropotkin had in mind. On the other hand, recognizing this kind of cooperation is important for those of use who have a tendency to think that cooperation is good in itself. Cooperation is desirable only to the extent that the goals it serves are desirable.

most people involved in literary study as well. It is an idealistic vision. But the mere fact that some of us favor it over hierarchical and violent alternatives may indicate that it is at least not an option excluded by our evolutionary history. Of course, such a condition will not be produced by literary narrative alone. Indeed, literary narrative sometimes contributes to belligerence. However, if we establish eudaimonia as our ideal—rather than the cultivation of individual superiority in wealth and power—then we will of necessity have a different vision of what constitutes value, not only for the arts, but for life generally.

13.5 Conclusion

Any discussion of literary narrative and evolutionary adaptation must begin from the distinction between mechanisms and functions. Specifically, it must consider what evolutionary mechanisms manifest themselves in literature, before considering the functions of such mechanisms. In connection with this, it is sometimes valuable to limit the concept of *by-products* to features that involve the same genetic sources as an adaptation, but produce distinct mechanisms. We may distinguish by-products in this sense from "adaptation-derived" features or practices, which result from interactions among the usual mechanisms of adapted systems. Such adaptation-derived features or practices may make a distinctive contribution to selective fitness without involving any further genetic changes. We may refer to this as "ancillary adaptation," distinguishing it from "strict adaptation," which involves the spread of an adaptive genetic mutation.

Universal properties of narrative are the key properties that might be open to an adaptationist account of the sort that has become so popular among Literary Darwinists today. However, as we isolate these properties, it seems that they are fully explained by the usual mechanisms of human cognitive architecture. In other words, they do not seem to require any additional account in terms of adaptation. Rather, they are adaptation-derived. There is some evidence that literature may have ancillary adaptational consequences. On the other hand, it is difficult to say whether these consequences are distinctive of literature or are shared by many practices.

These conclusions are likely to disturb literary theorists and critics. That is because such theorists and critics often see adaptation as a key value—even a value with possible monetary consequences. This is misguided in two ways. First, assertions that art is adaptation are unlikely to produce arts funding. Second, this accepts and even promulgates the equation of value with status productivity. It is more in keeping with the values cultivated by literature itself to stress the intrinsic worth of eudaimonia. Indeed, the greatest value of literature almost certainly does not lie in its conformity with the contingent dictates of adaptation. Its real worth may be found in the ways it can affect our emotional responses and, in that way, perhaps even alter just what is adaptive.

References

Cacioppo, J. & Patrick, W. [2008] *Loneliness: Human Nature and the Need for Social Connection* (Norton, New York).

Carroll, J. [2004] "Pinker, Dickens, and the Functions of Literature," in *Literary Darwinism: Evolution, Human Nature, and Literature* (Routledge, New York) pp. 63-68.

Darwin, C. [1872] *The Origin of Species*, 6th ed. (Murray, London).

Deresiewicz, W. [2009] "Adaptation," *The Nation* (June 8), 26-31.

Gould, S. J. [2002] *The Structure of Evolutionary Theory* (Harvard University Press, Cambridge).

Hatfield, E., Cacioppo, J. & Rapson, R. [1994] *Emotional Contagion* (Cambridge University Press, Cambridge).

Hogan, P. [2003] *The Mind and Its Stories: Narrative Universals and Human Emotion* (Cambridge University Press, Cambridge).

Hogan, P. [in press] *Affective Narratology: The Emotional Structure of Stories* (University of Nebraska Press, Lincoln).

Kropotkin, P. [(1902) 1989] *Mutual Aid* (Black Rose Books, Montreal).

Lam, L. [2008] "Human History: A Science Matter," in *Science Matters: Humanities as Complex Systems*, eds. Burguete, M. & Lam, L. (World Scientific, Hackensack, NJ) pp. 234-254.

Marcus, G. & Fisher, S. [in press] "Genes and Language," in *The Cambridge Encyclopedia of the Language Sciences*, ed. Hogan, P. (Cambridge University Press, Cambridge).

Nussbaum, M. [2001] *Upheavals of Thought: The Intelligence of Emotions* (Cambridge University Press, Cambridge).

Oatley, K. [2009] "Communications to self and others: Emotional experience and its skills," *Emotion Review* **1**, 206-213.

Pinker, S. [1997] *How the Mind Works* (Norton, New York).

Scarry, E. [1999] *Dreaming by the Book* (Farrar, Straus and Giroux, New York).
Tooby, J. & Cosmides, L. [1992] "Cognitive adaptations for social exchange," in *The Adapted Mind: Evolutionary Psychology and the Generation of Culture*, eds. Barkow, J., Cosmides, L. & Tooby, J. (Oxford University Press, New York) pp. 163-228.
Tooby, J. & Cosmides, L. [2001] "Does beauty build adapted minds? Toward an evolutionary theory of aesthetics, fiction and the Arts," *SubStance* **30**, 6-27.
Wong, R. [2000] *Motivation: A Biobehavioural Approach* (Cambridge University Press, Cambridge).
Woolf, V. [(1925) 1981] *Mrs. Dalloway* (Harcourt Brace, San Diego).
Zunshine, L. [2006] *Why We Read Fiction: Theory of Mind and the Novel* (Ohio State University Press, Columbus).

14

Emotion, Cognition and Aesthetic Form in Vishal Bhardwaj's *Omkara* and Shakespeare's *Othello*

Lalita P. Hogan

Discussion in this chapter uses insights drawn from cognitive theories of emotion to illuminate aspects of Shakespeare's *Othello* in relation to its 2006 Hindi film adaptation, *Omkara*, by Vishal Bhardwaj. The first section draws attention to the common motif of sexual jealousy and domestic violence, contextualizing it broadly, by drawing attention to how readers read stories in relation to their own private "stories," and Martha Nussbaum's notion that literary sensibility cultivates compassionate imagination. This context is central to considering the role of emotion in literature and film. The second section provides a brief introduction to the Appraisal Theory of emotion, explaining how emotions are elicited in relation to various types of cognitive evaluation of events and environmental factors, as to their desirability and undesirability, emphasizing the central role played by imagination. It is in this way that representational reality can have real emotional impact, even though the people and the situations are unreal. The third section provides details about plotlines of *Othello* and *Omkara*, highlighting main characters, events and relationship configurations. The fourth section takes up a discussion of the *law of situational meaning*, comparing the film and the play, with focus on the central conceptual metaphor of stealing and thievery used in connection with socially prohibited marriage. The fifth section focuses on the role played by visualization of the object, the gift of love, the loss and recovery of which aligns aesthetic form to the *law of apparent reality*. The sixth and final section, appropriately, shows how the emotion *law of closure* organizes stage, screen and story time, leading inexorably to the tragic *denouement* in the play and the film.

14.1 Reading Love and Anger

The following discussion of cinematic and dramaturgical emotion in Vishal Bhardwaj's *Omkara* [2006], a Hindi film adaptation of Shakespeare's *Othello,* draws on recent research done on emotion by cognitive psychologists such as Nico Frijda [1986, 2007], Paul Ekman [Ekman & Davidson, 1994], Richard Lazarus, Andrew Ortony [Ortony et al, 1988], and many others. My specific focus is on laws of emotion as they are systematically formulated by Nico Frijda and are included by Ed Tan in his overview essay in "Emotion, Art, and the Humanities," in the *Handbook of Emotions* (now available in its third edition). As summed up by Tan, Frijda's basic definition, "emotion is a system for the realization of an individual's concerns" has "control precedence." Since agreement on a workable definition of emotion is hard to find, Nico Frijda's sense that control precedence makes this definition "irresistible" [Tan, 2000, p. 118] in its ability to account for real life as well fictional information on emotion, augurs well for studying emotion in narrative.

14.1.1 *Imagination and Emotion*

In real life, as in drama and film, imagination plays a significant role in emotion elicitation. Through various kinds of diegesis (fictional and symbolic representation, as opposed to exegesis, or expository discussion), that make a story world 'real' to our concerns and imaginings, fictions elicit emotions and such elicitation is central to attentional focus, interest, and the overall experience of reading books and viewing films. The role of imagination for emotion elicitation becomes central in the play and the film because in each the protagonist only 'imagines' his wife to be unfaithful. The question, therefore, is what induces belief and what allows the subjects to draw inferences about a disturbing 'truth.'

<u>Language and belief</u>

Here, my intention is to draw attention to various verbal practices that are part of the human tendency to think in metaphors, metaphors that elicit beliefs about one's relation to the environment: social, familial and natural. Not only do conceptual metaphors induce belief; they also lead to inference about what is "true," as they influence our normative decisions about what is "reasonable" and "natural." For instance, it is implicitly accepted as a universally acknowledged "truth" that a man can become angry beyond reason if he thinks his wife cheats on him. Subjectively perceived human truths are not like scientifically proven truths; yet, once something is construed by the mind as "true," the impact on emotion (and action) is irreversible and predictable. Even though, typically, husbands in the 21st century do not consider their wives as their possession and property, a response of extreme anger at the news of a wife's infidelity is still portrayed in films and other forms of fiction in these terms. Concomitantly, putting up with the infidelity of one's wife would be considered unmanly by many in today's world. Heroic violence through some sort of vengeance may not be considered normal, but most will say they "understand."

14.1.2 *Marital Infidelity and Manly Anger: Varying Norms*

A contrary norm is presented not only famously by James Joyce, in his epic novel *Ulysses,* with Leopold Bloom spending his day evading, and even facilitating his wife's infidelity, but also, among other works of literature, by Isaac Bashevis Singer in his seminal short story, "Gimpel the Fool" [Singer, 1985, pp. 3-21]. In this story, Gimpel looks the other way when he knows his wife is sleeping with another man. As Gimpel puts it, "Another in my place would have made an uproar, and enough noise to rouse the whole town, but the thought occurred to me that I might wake the child. A little thing like that—why frighten a little swallow, I thought" [1985, p. 11]. In teaching this story nearly every semester in a general education course on International Literature, I find out that during the first segment of class discussion a vast majority of students are disdainful and consider Gimpel unmanly. They say they hate him, find him annoying and even get angry with him, while a tiny minority appreciates Gimpel's unmanliness in a world that is too manly for their own comfort level. They like his sensitivity, even admire him for putting fatherhood first, fatherhood of a child who is biologically not his. Progressing from shallow reading, my students eventually graduate to reading Singer's story for what it is, to pass the course, or to really cultivate what Martha Nussbaum calls "civic imagination," or "compassionate imagination" [1997, pp. 85-95].

14.1.3 *How Reading Is Influenced by Private "Stories"*

It must be pointed out, though, that the shallow reading habit is not atypical or random; it follows a psychological law whereby most readers read other people's stories narrowly in relation to their own private "stories." In the case of "Gimpel the Fool," my students perhaps read it in relation to their private stories of shame, various kinds of hurt and abuse, perhaps their not having been able to "stand up" for themselves, or having been ridiculed by their "manly" peers in high school, as Gimpel is constantly ridiculed.

1. <u>Shallow reading and deep reading</u>

On the subject of shallow understanding, I refer to Roger Schank's idea that "understanding" a story means that a reader/listener or a viewer maps it onto his own story. For Schank "understanding means finding a story you already know" and interpreting what you read/listen in relation to that alone [Schank, 1990, p. 71]. Expectedly, Schank thinks the readings of *Breakfast Club* he collected for his research present samples of varying degrees and kinds of "shallow" understanding [1990, pp. 56-83]. On the basis of his findings, he is inclined to believe that shallow understanding is the only understanding people generally have of other people's stories. Contrarily, the noted cognitive philosopher, Martha Nussbaum, believes in the possibility of the Aristotelian ideal: watching drama (and reading stories) for the *catharsis* of pity and terror. From this perspective, one can assume that understanding (and experiencing) literature leads to empathy. This type of reading is not shallow and is not limited to mapping one's own stories onto others, but results in "cultivating the powers of imagination" [Nussbaum, 1997, p. 85] when, for instance, a classical play "invites its readers to know and see more than the unseeing characters" [1997, p. 87].

2. Beyond shallow reading, changing private "stories"

When my students put aside their individual variants of shallow reading (readings that those of us who teach literature know to be infinitely valuable), and move towards what Martha Nussbaum would call a reading process that engages the imagination, they *see* what is hidden from Gimpel's wife, the townspeople, even Gimpel's grown up children. In other words, they *see* what the "unseeing characters" do not *see*. In this process, they realize that Singer inverts gender roles and, by design, attributes to Gimpel all the traditional virtues of a woman, such as her ability to keep peace in the home at great cost to herself; tolerate her husband's infidelities; by sheer instinct attach the highest micro-goal priority to protecting a new born baby's sleep. If Gimpel is a fool, then every proverbially "good" woman is a fool. For Singer, who lived through the holocaust and experienced the most horrific manifestations of aggressive masculinity, *Gimpel is what a man should be, a fool*; in so far as that is what it takes to act without violence even when there is provocation for violence. Provocation for violence does not justify violence.

3. Reading for empathy

Since violence is pervasively associated with self assertion and non-violence with self-effacement, my students' shallow reading is relevant for their entire reading experience. Most importantly, in the process of reading for empathy and seeing what other characters do not see, their own "story" is changed. They forgive themselves, by extending empathy to their own self. Perhaps they no longer feel shame for having been fooled, for not having been able to "stand up" for themselves, when, in most cases, doing so might have involved them in violence of some sort. This rite of passage from one kind of reading to another kind leads my students to declare, without any fear of contradicting themselves, that they "love" Singer's "Gimpel the Fool," and the story invariably ranks among the top three in their "best choice" list at the end of the semester.

4. Masculinity, violence and righting wrongs

In contrast to Singer's Gimpel, Omkara and Othello are manly men, heroes of tragedies, not town fools. They are protagonists in a canonical play and a mainstream film. Ajay Devgan, who plays Omkara, is primarily an action hero; the ballad at the outset of Bhardwaj's film celebrates him as the greatest warrior (sabse bad ladaya re). The question readers and viewers ask in relation to these texts is why can't Omkara/Othello figure out that his deadly enemy, posing as his friend, is lying to him? They rarely ask, given the provocation, should Othello/Omkara have reacted the way he did? In part this is due to a greater emotive focus in the original play on Othello's relation to Iago, and less on his relation to Desdemona. That is why structural irony in *Othello* centers primarily on cognition associated with the discrepancy between what the reader/spectator knows, *Iago is a villain,* and what the tragic protagonist and other characters think, *Iago is honest.* This discrepancy guarantees a murder and suicide ending, while bridging of the knowledge gap at the end assures separation of "truth" from "falsehood," friend from foe. In general, our acceptance of stories involving violence, and our tendency to enjoy them, *does have* something to do with normalization of various kinds of violence depicted in literature and film. With regard *Othello* and *Omkara,* there is something else going on that goes beyond the question of gender, though gendered nature of violence (associated with male behavior) and non-violence (associated with emotional reactions of females) remains a constant.

5. Not appraising trusted friend as enemy

With regard to the Othello/Omkara story, or other texts like these, however, narrative energy and aesthetic emotion, as well as attentional focus, is cognitively anchored more in the protagonist's inability to tell friend from foe, separate folly from wisdom, truth from falsehood. In light of the high adaptive importance of being able to make these cognitive distinctions, the readers/viewers interest, concern, investment is intensified by the catastrophic, irreversible consequences of this inability, or temporary incapacitation due to a trusted friend's sinister deceit.

Even though there is a marked difference between Shakespeare's occlusion of Othello's direct involvement in war killings and Vishal Bhardwaj's comprehensive visual coverage of Omkara's engagement in political violence, both texts involve the protagonists into making bad cognitive judgments about who is enemy and who is friend; what is true and what is false; whom they should trust and whom they should distrust. This activates a reader's generalized, free floating concern and anxiety about these issues.

14.2 Laws of Emotion, Appraisal and Imagination

Drawing on appraisal theory of emotion, my specific literary and film analyses in the following discussion refer to three laws of emotion: the *law of situational meaning*, the *law of apparent reality* and the *law of closure,* following Nico Frijda's formulation of the laws of emotion and their relation to imagination and aesthetic form. While I will limit myself to these three laws of emotion, other laws, such as the *law of change*, the *law of hedonic asymmetry*, the *law of concern,* as defined by Frijda, can allow us to productively re-read a great deal of literature in light of theories of emotion developed in cognitive science. Such inter-disciplinary readings can also enhance and reinforce the scientific project of studying emotion.

14.2.1 *Relation between Appraisal and Laws of Emotion*

For instance, the "law of change" posits that emotions are not elicited by the presence of "favorable and unfavorable conditions," but "actual or expected changes" in favorable and unfavorable conditions [Frijda, 2007, p.10]. This law, quite clearly, articulates the idea of reversal of fortunes (from favorable to unfavorable, and/or unfavorable to favorable) in narrative. Aristotle called this pattern *peripeteia.* The "law of hedonic asymmetry" has to do with "comparative feeling," a "stern and bitter law" that is based on how the emotion system becomes habituated to established modes of pleasure, so that change, even when it might disrupt domestic order, or any other form of an ordered life, is necessary to

render pleasure [2007, pp. 12-13]. The "law of concern" is an overarching law according to which emotions "arise in response to events that are important to an individual's concerns" [2007, p.7].
These concerns can be narrowly personal, or broader, having to do with larger social anxieties about economy, justice, equal distribution of resources, issues that affect women and minorities in societies. In addition, the concerns can be real, but they can also be imaginary, the kind we find in *Othello/Omkara*: the concern about a woman's nature.

Beliefs and appraisal

Essentialist beliefs about womankind (as about races) stipulate a "truth" about what a woman's nature is: changeable, something one cannot build one's trust on. Brabantio famously articulates this belief (about his own daughter) in Shakespeare's play. In the Indic context, reference is made to scriptures that postulate this notion, and an incisive critique of *triya-charitra* (woman's nature) is eloquently articulated by Indu in *Omkara,* when she confronts the suspicious Omkara on the terrace, and tells him to stop the wedding if he has the slightest doubt about Dolly's virtue and chastity.

In a way, Shakespeare's play shows that this essentialist belief about womankind is what Iago and Othello share. Overcome with emotion, when Othello says to Iago, "I am bound to thee forever" (3.3.217),[1] and Iago says "I am your own forever" (3.3.482), an uncanny utterance that closes this fraught scene, marks a strange Vulcan mind-meld of two men, who, as individuals, could not be more unlike each other. It is at least in part due to this shared belief that Iago is able to alter Othello's "constant, loving, noble," and stable nature—to which Iago himself attests (2.1.287-288)—so that "the Moor" mimics a fabricated inconstancy, changeableness, generally attributed to a woman and particularly to his beloved wife. In contrast, the play shows Desdemona to be constant—unchangeable almost to a fault. This is done to make visible, to question, and to invalidate a belief rooted in men's psyches,

[1] All direct quotations from Shakespeare's play(s), using the Arden edition of *Othello* and the Signet edition of *Romeo and Juliet* will appear in the following, standard citation format throughout: (Act. Scene. Line(s)) from [Shakespeare, (1604) 1999] and [Shakespeare, (1596) 1987].

implanted by perverse and persistent gender ideologies. Emotion laws, as Frijda maintains and as the film and play demonstrate, involve multi-level information processing, as well as multi-level cognitive appraisal: voluntary and involuntary, instantaneous and deliberate.

14.2.2 *Basics of Appraisal and Early Modern Theory of Humors*

The basic tenet of Appraisal Theory is that emotion is not elicited by life events by themselves, but by how events are appraised in terms of their impact on goals, purposes and desires, on the basis of aversiveness and attractiveness of effects, affects and possibilities: their being either desirable or undesirable. In literary studies, this research has had the most distinctive impact on early modern studies and Shakespeare studies. This is due to the fact that students in this area have for years been researching Galen, Hippocrates and others to understand the literature and moral philosophy of this period. They know that the psychological notion of character used in this period was partially based on the physiological theory of four humors and body fluids, their impact on mood, and mood congruent thinking that can lead to certain behavior patterns and action outcomes. Thus, character prototypes—the choleric (prone to anger, due to excess of yellow bile in the body), phlegmatic (dull and unresponsive, due to excess of phlegm in the body) and sanguine (high spirited, due to excess of blood in the body)—provided systematic material for comedy and social satire. The melancholy humor (caused by excess of black bile in the body) did even more. The impact of melancholy on mood contributed to the idea of a certain type of a tragic character (such as Hamlet) and also a grand theory about poetic inspiration. This physiology, as physiology, has become outdated, but the early modern dramaturgy of embodied consciousness (of characters) is not bound by the medical dogma of a certain period. Hence, individual works of drama of this period and their modern adaptations lend themselves to being studied as mental models of emotion systems, as they are conceived today. The findings of cognitive psychologist like Nico Frijda and many others are, on the one hand, based on empirical research; on the other hand, our study of literature and film impacts this research, as it is impacted by it.

14.3 *Othello* and *Omkara*: The Main Storyline

As is well known, *Othello* is the story of a Venetian general of Moorish descent, who marries a girl from a noble family against the wishes of her father. While serving the Venetian state in Cyprus, he is made jealous due to false reports and insinuations about his wife's infidelity by Iago, his assistant. In response to a sense of wrong done to him, leading to extreme anger, he murders his wife. A few minutes later he commits suicide when he finds out that he was tricked by the villainy of someone he trusted and that he killed his wife in error. Vishal Bhardwaj's Hindi film *Omkara,* an adaptation of *Othello,* as pointed out above, follows the same storyline, with a changed ending. While Othello is an African living among the Venetians, Omkara is a half-caste; his mother is low caste and father is a high caste Brahmin. Also, while Othello's military career is nobly devoted to service of the Venetian state, which he defends against the Turks; Omkara is the fearless leader of a gang that carries out political crimes for local politicians. The context for this film is the violent political culture of a postcolonial nation, with a great deal of focus on rival political parties, their covert armies in exclusive communities, with their own codes of duty, revenge and honor, rites and laws, as well as hierarchies of subordinates and superiors.

In the backdrop of an antiquated, but still practiced heroic age ethos of strong armed violence juxtaposed with modernity and its many gadgets, such as cell phones, trains, cars, guns, Omkara interrupts the nuptials of Dolly Mishra, the woman he loves and who loves him. While in Shakespeare's play, Iago's villainy is almost motiveless (with his stated "reason" that Othello made Cassio his captain in place of Iago being presented as no more than a very thin excuse), Vishal Bhardwaj's Ishwar Langda Tyagi's malevolence is motivated. He feels deeply wronged because Omkara chooses a less heroic figure, a mere college student, Kesu (Keshav Uppadhyay), as his bahu-bali, the strong armed enforcer, and that too after Tyagi has single handedly helped with the abduction of Dolly Mishra. As one would expect, Langda Tyagi, nicknamed so because of the limp, carries out his revenge by making Omkara believe that Dolly has had an affair with Kesu, because they studied at the same college and have much in common, while Omkara and Dolly come from very different worlds and have nothing in common.

14.3.1 *Make-Believe Reality and Emotion in* **Othello** *and* **Omkara**

It is this make-believe element that aligns plot transformations and character consciousness with the appraisal theory of emotion. Truth status of an imagined realty is initiated and maintained by one very determined person who functions like an invisible author of the embedded plot of domestic violence in the backdrop of a larger plot of political violence in the play and the film. In a sense, the play and film present mental models of a world where some kind of strong and insistent propaganda, passed off as "truth," will lead people to appraise events and persons in such a way as to elicit anger, leading to action outcomes of violence.

14.3.2 *Metaphor, and Appraisal of Women as Objects*

More significant for the following discussion of domestic violence in the context of heroic virtue and a heroic ethos, are appraisal models that are culturally and cross culturally derived, such as concept-ualizations of women as objects that can be stolen, territories that can be invaded, usurped, or rights and claims that can be appropriated by someone else. Once this cognition has been made, it follows that something must be done to negate this reality, to remove the cause to serve a cause, as this idea is articulated in Othello's memorable words: "It is the cause, it is the cause, my soul!" (5.2.1). I refer here to conceptual metaphor theory that allows for "understanding one conceptual domain in terms of another conceptual domain" [Kövecses, 2002, p. 4], such as understanding of life (target domain) as a journey (source domain), desire as fire or hunger, argument as war. Source domains are usually the imagery and metaphor, though not always, and target domains are usually, not always, abstract entities [Kövecses, 2002, pp. 3-27]. With regard to the central story elements in *Othello* (and emotions they are anchored in), a variety of conceptual metaphors come into play where many domains are mapped onto each other. With dizzying speed cognitive appraisals elicit emotion, and emotion leads to action and mood congruent cognition, resulting in the deaths of Othello, Desdemona (and others)—"the tragic loading of this bed" (5.2.361) that Lodovico invites not only Iago, "the Spartan dog" (5.2.359), to "look on" (361), but the reader and the spectator as well.

14.4 The Law of Situational Meaning: The Foul Thief in *Othello* and *Omkara*

Defining the *law of situational meaning*, Nico Frijda says that if we "input some event linked to a particular kind of meaning, out comes an emotion of a particular kind of meaning. This is the law of situational meaning in its simplest form" [2007, p. 5]. Frijda turns to films and stories for sample events of this sort, which elicit tearful emotions, through such prototypical themes such as "separation-reunion, justice in jeopardy," or virtue under siege, and so forth [2007, p.7]. In *Othello,* the "virtue under siege" and "justice in jeopardy" emotion-input themes apply to Desdemona, with regard, first, to her unawareness of evil. Incurring her father's wrath, Desdemona married Othello, not knowing that someone close to her husband is lurking around the corners to turn the "luscious," sweet and juicy fruit of love into "as acerb [bitter tasting] as colquintida [bitter apple]" (1.3.349-50). The reader knows Iago is anticipating "hell and night" to bring "this monstrous birth to the world's light" (1.3.401-402), envisioning *that* as the final outcome of Desdemona's fairytale romance and marriage. This demonic anticipation contrasts with Desdemona's trust in Iago as she kneels: "Good friend, go to him, for, by this light of heaven,/I know not how I lost him. Here I kneel" (4.2.152-53). Dolly does not have as many direct scenes with Langda Tyagi (Iago); there is nothing to show that she trusts him, or that she thinks much of him. This changes the *situational meaning* in the Hindi film.

14.4.1 *Romantic Love as a Figure for Democracy*

The micro-genres of *virtue under siege* and *justice in jeopardy*, thus, are not the *input* emotion themes in *Omkara* so far as the characters are concerned. The input emotion themes of justice in jeopardy and virtue besieged relate to the public being duped, badly used, with their trust abused by politicians. Clearly, Bhardwaj's adaptation of *Othello* invents its own, rather broadly defined, conceptual metaphor in the form of narrative allegory, where electoral politics in a democracy that values people's choice to elect their officers is equated with family politics, visually articulated as *democracy is romantic love and marriage*. The problematic construal, or appraisal, questioned by the film is that in both (love and politics), one has to *fight for the right to choose, even if the fight involves violence*. Friar Lawrence's comment to Romeo, that "these violent delights have violent ends/And in their triumph die" (2.4.9-10) is forgotten. When he tries to find in the tangled web of a love story a straight path to civic harmony in Verona, the caution is left aside by Friar Lawrence himself in *Romeo and Juliet*—as the good man plays with sleep inducing potions to bring about a feigned death. In *Omkara,* a girl desiring to "elect," to choose whom she loves, has to rely on the very same gangsters who are in the employ of the electoral candidate who thinks people's best interests lie in choosing him, and his party. From this perspective, love's narcissism, its idolatory, is aligned with war and politics. From another perspective, though, love is presented as antithetical to war and politics. The paradoxical nature of love implicates it in violence, and also defines it against violence.

14.4.2 *Appraising Love as Possession and Its Loss as Theft*

With the focus on power relations between men and women, the situational meaning foregrounds the question of violence, and the central conceptual metaphor that organizes discourse at this stage is, LOVE IS POSSESSION. The emergence of separate sphere ideology in early modern England, as well as the pervasiveness of the "separate spheres" notion of duty, obligation and rights for men and women in all traditional societies, including many contemporary South Asian societies, contribute to a configuration of associated conceptual metaphors such as, WOMEN ARE OBJECTS. Figuratively, through extension, elaboration and combination of source domains ("clearly delineated concepts") and target domains ("less clearly delineated concepts") [Kövecses, 2002, p.15], the categorial identification of woman with the home, or even the identification of nation as mother, contributes to appraisal of behaviors, values and actions. It is in this way that the law of situational meaning in *Othello* and *Omkara* engenders the rhetoric of theft and stealing for cognitive processing of information about socially forbidden marriage. The poetics of the film and the dramaturgy of the play are organized by this rhetoric. In *Othello,* when the Duke of Venice advises him to make the best of "the mangled matter" of his daughter having married a "sooty bosomed" Moor, Brabantio, the irate father retorts: "So let the Turk us of Cyprus beguile?/It touchés us not as long as we can smile" (1.3.210-111), equating Desdemona with Cyprus and accusing the Duke of a double standard. If it is within the rights of the Duke to send a large army to prevent "the Turk" from capturing Cyprus, why must Brabantio not demand a legal remedy for the loss of his daughter to a Moor? The binary opposition in the two roles Othello plays—as the "foul thief" who *stole* Desdemona and the brave general who will prevent Cyprus from being *stolen* by the Turk from the Venetian state—foregrounds the rhetoric of irony based in the *law of situational meaning*.

When Brabantio feels sure that the injury done to him can be rectified by Venetian law and he comes with his own men to "arrest" Othello (1.2.77-80), he accuses the "wheeling stranger" of having practiced "foul charms" to "enchant" his daughter. He charges that the Moor has "Abused her delicate youth with drugs and minerals" (1.2.74). Invoking a context not of witchcraft and sorcery, but of the burgeoning mercantile economy of early modern Europe, Iago refers to the marriage as the Moor having boarded (a Spanish) treasure ship (to *rob* or to gain legitimate access to the treasure) when he says; "Faith, he tonight hath boarded a land carrack" (1.2.50). Even Cassio, in his hyberbolic praise declares: "The riches of the ship is come to shore" (2.1.83), as he welcomes "the divine Desdemona" to Cyprus. Even as Cassio does so, Iago promises that Roderigo, Desdemona's jilted lover, will be able to *steal* her affections soon enough. In the opening scene when Roderigo and Iago wake up Brabantio in the middle of the night to tell him of the marriage, Iago says: "you have been robbed" (1.1.84). Later, Othello refers to his wife's "stolen hours of lust" (3.3.341-345). He wishes he had never known about them, and concludes with an enigmatic, "Let him not know't, and he is not robbed at all" (346).

Metaphor of theft and cognitive appraisal

In other words, the so called *robber* of Brabantio's daughter is now being cognitively primed to appraise that he has been *robbed*, though he wishes he knew nothing, so he would not experience the emotions he is experiencing and would not have to take action. In terms of metaphoric cognition, it is the villainy of language that Iago embodies; he *is* a kind of language. Linked to a conceptual metaphor, then, that leads to cognitive processing of information, and making value judgments, the inter-racial marriage is appraised in terms of cultural scripts for magic and its impact on emotion, material economy and territorial politics. As the discourse develops, the central, metaphorical concept that love is possession is rhetorically extended and elaborated to infer that woman (as daughter or wife) is territory, property and possession. If a woman is an object, it is inconceivable that she has a will, and if she is virtuous she has to be obedient, and would not go against the wills of others. Thus, if she is stolen and magically enchanted to disobey, strong armed tactics must be used to get back what one has an entitlement to.

With a different emphasis, the metaphor of theft, stealing, re-appropriation, redistribution of a value embodied in Dolly's character, is foregrounded in song, visual image and narrative. In *Omkara,* Dolly Mishra invites Omkara to interrupt her nuptials and abduct (*steal*) her. As a parallel to Iago's false promise to Roderigo that he can gain access to Desdemona once she is tired of the Moor, Langda Tyagi gives assurance that Rajju only has to wait till he gets his *lost* bride back as Omkara's adulterous wife. Though Dolly Mishra is not compared to territory as Desdemona is, she is emblematizes in terms of the symbolic economy of value, privilege and access that only middle class urban elite in India have, access to education, sophistication, free choice, even access to a free and fair electoral process. Because of her fairer complexion, Dolly is also appraised as racially or ethnically superior to the dark colored Omkara. Just as Desdemona's father appeals to the Duke of Venice to intervene, Dolly's father appeals the private court of Bhaisaheb, who is the politician whom all these people serve. In front of this dignifiedly corrupt politician, the injured father weeps and asks; how can it be allowed that his daughter should be taken away, or given to, this daitya: this demon. In Indian mythology, daityas are the prototypical enemies of devas: the gods. Ironically, Dolly's father, Raghunath Mishra, a respected lawyer, is the one who protects the daityas, saving them from being proven guilty (often of murder) in courts of law. He empowers and enables them to "steal" elections and, thus, taint safe and fair electoral processes. His daughter's freewill, however, is not something he can control because someone more strong-armed now guards her enfranchisement. It is as if, metaphorically, the population of this region has given their consent for the elections to be hijacked in favor of this particular political party, and this particular candidate.

14.4.3 *Malevolence: Stealing the Gift of Love*

In *Othello,* the foul thief is ultimately not Othello, but Iago, who has his wife steal Desdemona's handkerchief so that this object can be manipulated to serve as a proof of infidelity. In *Omkara*, Langda Tyagi's wife, Indu, steals a heavy, bejeweled, waist ornament. There is some ambiguity about whether she steals it on her own, or if she is asked by her husband to steal it. Indu's and Emelia's theft, though accidental and unthinking, is heavily implicated in the surrounding malevolence. As women, and as free agents, all their actions are not in their conscious control and can be suggested by the overall rhetoric of stealing and theft, robbing and being robbed. Adaptive significance of the primitive fear of being robbed of home, goods and life, a concern that motivated groups to enter into social contract (origin of governments and nation states)—this subtext of evolutionary history may imbue everyday concerns about trust, about being able to separate friend from enemy, truth from falsehood, with greater emotion eliciting potential.

While in *Othello*, this question is posed as Othello's and everyone else's misplaced trust in the proverbially "honest" Iago, Bhardwaj highlights an atmosphere of political blindness, confusion, misinformation about social issues, power-grabbing acquisitiveness of electoral candidates. The personal choice that Dolly has made is clear and authentic; her vision and feeling sustain her even in the moment of her brutal death. Suggested partially by the primary colors she wears throughout—blue, white, yellow, red, and green—her clarity of feeling in the film is surrounded by environmental fog, clutter of objects, dust, smoke, suggesting false alternatives posing as real choices for the populace. Thus, Dolly, though not virtue incarnate like Desdemona, is, in the Indian context (Hindu and Buddhist) like the proverbial lotus, blooming in a muddy pond, untouched by the mud. The visual idealization equates her with democracy, such as might have been envisioned by Mahatma Gandhi (as well as other makers of an independent nation).

14.5 Appraisal and the Law of Apparent Reality

Emotions are "subject to the law of apparent reality," because they are elicited by "events with meanings perceived as real" [Frijda, 2007, p. 8]. That is, "events that are taken to be real elicit emotions" (p. 8). Frijda refers to the "vividness effect" discussed in journalism and social psychology, about how news photos have more impact than expository discussions and abstract reporting [2007, p. 9]. Here again, Frijda makes reference to "objects of art—drama, novels, paintings" having "strong emotional impact while one is perfectly aware that they are fictional" [2007, p.10], because they appear real to the imagination due to sensory perception.

14.5.1 *Theft and Its Imagined Appraisal in* Othello *and* Omkara

The central conceptual metaphor, as referred to above, LOVE IS POSSESSION, when extended through a concrete object, animates an entire dramaturgical design through an insistent visual focus on an object (as a stage prop and as part of emotively motivated *mise en scene*). This object is lost and stolen in the *complication* and *rising action* phases of plot development and stage action; it appears on stage (and screen) as a proof of the wife's infidelity in the *climax* and *falling action* phase of plot development and stage action, thickening the veil of *agnoia* (ignorance). It leads to belated *anagnorisis* (*recognition*) at the end when it is brought back into visual focus both as proof of the about-to-be-murdered wife's guilt, evidence that firms up the murderous resolve, and evidence that exonerates the just murdered wife, posthumously proving her innocence.

In this last part, the metonymic object (representing the woman's sexual body) and a metaphoric object (representing how it figures in male imagination) organizes *denouement* of the play and the film, though Vishal Bhardwaj substitutes a heavy piece of jewelry for the little handkerchief in Shakespeare's play. It is this object that visually anchors frames of changing inferences about belief and unbelief, trust and distrust, guilt and innocence, purity and impurity.

14.5.2 *Discourse of Slander, Appraisal, and Emotion*

As slanders and untruths spread through verbal exchanges between people, the object changes hands, and spreads like some sort of a virus. Because it is the contagion of slander, it is inescapable. To this effect, Iago refers to his lies as "medicine" and "poison," as he gloats thus: "The Moor already changes with my poison" (3.3.327); "Work on,/My medicine work!" (4.1.44-45). Othello's haunting image of the raven flying over a house of death, his metaphoric representation of the troubling thought (truth, and belief) induced by how he appraises Iago's slanderous lies about Desdemona, is translated into montage centered on fear and foreboding in Bhardwaj's film.

Specifically, Othello says: "O, it comes over my memory/As doth the raven o'er the infectious house/Boding to all—he had my handkerchief" (4.1.19-22). These lines, in my view, are the source for one of the central visual motifs in *Omkara,* involving a snake falling from the mouth of a bird into a pot filled with unguents for the bridal bath. The bird, flying over the blue skies, uttering horrible bird-shrieks (perhaps the bird is gloating at his triumph, as Iago is) disrupts pastoral wedding preparations surrounded by ambient sound, song, dance, carried on by the *unknowing* rural community of women. Dolly screams and runs inside the house where she is comforted primarily by Indu, though Langda is present to protect his investment, the poisonous lie, by not allowing the conversation between Indu and Dolly reach a point of untimely revelation. Insertion of this highly emotive *mise-en-scene* and montage, the shrieking, first of the bird, then of Dolly, contributes not only to the vividness effect in connection with a foreboding omen, it draws further attention to inferences about "truth" that the human mind-brain relies on for cognitive appraisal of events as to their being desirable or undesirable. Even though Omkara is going through with his wedding to Dolly (may be because the event has become publically and in a sense politically over-subscribed, as Dolly's wedding to Rajan had been at the beginning), the groom's body language sends threatening signals to the bride. She cannot envision her murder at his hands, but her mind is filled with fear. She says to Indu, "This is not at all the Omkara I knew" (yeh to voh Omkara hai he nahi); yet, this is the Omkara she is marrying.

14.5.3 *Discursive Circulation: The Stolen Gift of Love*

In *Othello,* the handkerchief is attributed a rich prehistory in vivid language after it is lost, and the purpose, again, is to induce a sense of reality that would elicit guilt, shame and foreboding in Desdemona. As a parallel to this imagery, the resplendent beauty of the waist ornament in *Omkara* is displayed through the dancing figure of Billoo Chaman Bahar, Kesu's whore, dancing for the BSF (border security force) soldiers, playing with their pistols, mingling the violent with the erotic in a Dionysian carnival. In the film, this is the moment of climax; the falling action phase is filled with the wedding, and the *denouement* with murders and suicides, in which the ornament figures as a poignant evidence of guilt and innocence. Prior to this, in the dance hall, an eye-line match is created between Langda's stupefied gaze and the dazzling ornament. It is clear that before he marks the inebriated Kitchloo,

Omkara's eye falls on what he infers as the debased gift of love, exchanged by his fiancé for a desultory dalliance with her supposed boyfriend who, as in the original play, gives it to his whore. The journey of this figure and its visual behavior is noteworthy for the activation of the law of apparent reality (Fig. 14.1).

The information about Omkara having seen the stolen family heirloom he had gifted to his wife to be, Dolly, is communicated only through a reaction shot of Langda's face. The link between the home and the brothel (as a parallel to Othello's enactment of a brothel scenario in his own home in 4.2) is further enhanced by Billoo flaunting this dazzling piece of jewelry over a black dress, while Dolly had worn it over a red dress in the privacy of the home, only for her own pleasure and that of Omkara. Indian brides wear red; hence, red is associated with socially sanctioned sexuality, while black, though not the color of death, is the color of mourning as well as of unbridled sexuality.

Fig. 14.1. *Upper left*: Indu steals the ornament and gives it to her husband. *Upper right*: Indu's husband, Langda Tyagi, gives it to Kesu to entice Billoo. *Lower left*: Kesu gifts it to Billoo. *Lower right*: Billoo performs at a specified venue, to lure Kitchloo.

The refrain of Billoo's song, in contrast to the licentious atmosphere of the room, speaks of single minded devotion to one person. It refers to the salt of love that has clung to the tongue. From the perspective of this song the conceptual metaphor, love is possession, is questioned. The language of the song appraises love in its own terms; love is love's own taste, sentimental and visceral, the taste that clings to the tongue like the taste of loyalty. As salt, in the Indian context, is a conceptual metaphor for loyalty, and here it alludes to Mahatma Gandhi's Salt March, an iconic event in India's independence movement, a politically emotive narrative of the home and the nation, along with all the expected ambiguities, is enacted in Billoo's song. At this juncture, the diffusion of "zuban pain laga, laga re/namak ishk ka/balam se manga, manga re namak ish ka" (stuck to my tongue today is, /Is the salt of love/[because]I demanded from my lover/Lover, the salt of love)—like that of some intoxicating fragrance, seeps into the smog and smoke of the mean streets outside. The ornament is for Billoo (not so for the viewer) a promise of Kesu's commitment of marriage, the "salt of love" she thinks she has given and received. Her figure is split between the speaker in the song lyric, and the courtesan-dancer who uses the art of her body and the magic of her voice to entice the soldiers, while her eye is on Omkara's target for today's murder: Kitchloo. The political and the personal are linked through strategically calibrated sound bridges between Langda and Omkara's vans arriving for the assassination, as the strains of music and song are heard from the street.

14.6 The Law of Closure: Putting out the Light

The law of closure refers to the self contained nature of an emotion, emotional reaction or an emotion episode, its absoluteness, characterized by Frijda in the following words "...emotions know no probabilities, no likelihoods. What they know they know. .... Your jealousy is certain. ....Your anger is certain" [2007, p. 16]. The closure of emotions shows itself in what people do, by the urge to act, and by the absoluteness and finality of the act itself. After he has decided that "she must die, else she'll betray more men," as mentioned earlier, Othello reflects: "It is the cause, it is the cause, my soul!/Let me not name it to you, you chaste stars/It is the cause" (5.2.1-3). The absoluteness of the law of closure that makes subjects "*...closed to the requirements of interests other than those of their own aims*" [Frijda, 2007, p.15] shows not only in Othello's resolve, but his cognitive aggrandizement of it.

Appraisal, emotion, and narrative causality

If love is possession, and wife is space that has been invaded and contaminated, it must be repossessed first through murder when the wife is thought to have challenged that idea and, later, through suicide when it is proven that she was innocent and the murdering husband has made an error ridden cognitive appraisal. The absoluteness of love and anger marks the end in the play and in the film. While Othello's final words, inadvertently, link him to the "turbanned Turk," identifying him as the enemy of the state, though his literal request to those who witness his death is that his services to the Venetian state should be remembered; Omkara too identifies himself as the enemy of the state, as well as of the home, an ingrown, deeply grafted enemy whom the state has created and whom the home has created through socialization of emotion that cannot escape the error of gender ideologies and appraisal models derived from them. In political terms the bride in red who lies dead on the sadly swaying swing is something larger, like a Polity that ought to have been saved. Beneath it, revealed and concealed by turns, is the white robed body of the bridegroom, who had always worn khaki, and/or red and black. His corpse is now an emblem of the white-robed politician who alternates between being a public servant and a criminal.

Thus, we have seen that the emotion law of closure necessitates the murder and suicide ending in both the play and the film because once an emotion has been elicited it will run its course. The inevitability of death, or fate, no doubt, follows the genre law of tragedy; yet, *the law of genre encapsulates the emotion law of closure*, showing how cognitive theory of emotion and aesthetic form meet at the point where science meets art. Along the way, various types of antecedent appraisal have been made by socially and culturally derived cognitive appraisal models, as discussed above.

Some of the models deal with essentialist beliefs about women (and men), some with what is right in love, war and politics, others pertain to expectations, rights, entitlements of each in love and marriage; still others are changeable according to the particular situation and lead to drawing (false) inferences from information provided by a person who is trusted, but who should not be trusted. Truth and falsehood have been mixed up to prepare this "poison," induce this, hopefully, cathartic delusion.

While Bhardwaj's film solicits an engagement in the Indian viewer that would enhance "civic imagination," as Martha Nussbaum describes the humanizing function of literature and film in a democracy, Shakespeare's play tells the story of a man who, in his own words, "loved not wisely, but too well," someone "not easily jealous, but, being wrought,/Perplexed in the extreme," and most importantly, the one who "Drops tears as fast as the Arabian trees/Their medicinal gum" (5.2.342-449). In the last two lines the reference is to myrrh trees, and by association to the substance that was among the gifts that the Magi brought for Jesus at his birth. Othello's vow to die upon a kiss (5.2.356) is a gesture of commitment to the "salt of love" (namak ishk ka) referred to in Billoo's song. The purifying efficacy of his tears may be another one of Othello's grand gestures; yet, it expresses a desire to set things right, to heal what has been harmed.

References

Bhardwaj, V. [2006] *Omkara* (Indian film).

Ekman, P. & Davidson. R. J. (eds.) [1994] *The Nature of Emotion: Fundamental Questions* (Oxford University Press, New York).

Frijda, N. H. [1986] *The Emotions* (Cambridge University Press, London).

Frijda, N. H. [2007] *The Laws of Emotion* (Lawrence Erlbaum Associates, London).

Kövecses, Z. [2002] *Metaphor: A Practical Guide* (Oxford University Press, London).

Nussbaum, M. C. [1997] "The narrative imagination," *Cultivating Humanity: A Classical Defense of Reform in Liberal Education* (Harvard University Press, London).

Ortony, A., Clore G. & Collins. A. [1988] *The Cognitive Structure of Emotions.* (Cambridge University Press, London).
Schank, R. C. [1990] "Understanding other people's stories," in *Tell Me a Story: Narrative and Intelligence* (Scribner, New York).
Singer, B. I. [1985] "Gimpel the fool," in *Gimpel the Fool and Other Stories* (Farrar, Srauss and Grioux, New York).
Shakespeare, W. [(1604) 1999] *Othello,* ed. Honigman, E. A. J. (Arden, London).
Shakespeare, W. [(1596) 1987] *Romeo and Juliet*, ed. Barnet, S. (Signet, New York).
Tan, E. [2000] "Emotion, Art, and the Humanities," in *Handbook of Emotions,* eds. Lewis, M. & Haviland-Jones, J. (Guilford, New York).

15

Tanbi Novels and *Fujoshi*: A New Romance for Young Chinese Women

Ting-Ting Wang

In the 1960s, *tanbi* comics and novels which focused on boy's love came into being in Japan and was popular among young women. They, both readers and writers of tanbi, were women and called themselves *fujoshi*. At the late 1980s, Japanese tanbi comics but not novels became available in China. It was not until the end of the 1990s that tanbi novels with Chinese authors appeared in novel websites in China. Between 2005 and 2006, it became a main style of net novels in China. In this chapter, we first introduce the history of tanbi subculture in Japan, the two main types of tanbi texts, and the writing and publishing of tanbi novels in China. We then address the questions on why tanbi subculture is well accepted in China, what desires of young women are reflected in it, and how it challenges the mainstream culture.

15.1 Introduction

In the 1960s, *tanbi* comic came into being in Japan, the subject of which is boy's love. Handsome young men and boys, beautiful love and hot sex are universal elements in tanbi comics and novels. Most of he readers and writers of tanbi are young women with ages from 15 to 30; they call themselves *fujoshi*. After the introduction of tanbi comics to China in the late 1980s, tanbi novels written by Chinese started to appear in novel websites in China at the end of the 1990s. In the subsequent several years, they have only a small number of readers and writers. However, between 2005 and 2006, tanbi novels suddenly began to have more and more readers and writers, and became a main style of net novels in China.

Subsequently, in October 2009, Snda Net Novel Co.[1] in Shanghai and the State Post Bureau of China issued five sets of stamps consisting of 64 pieces, on each of which was the portrait of a net writer. For each of the three novel websites owned by Snda—Jinjiang, Qidian and Hongxiutianxiang—there was 16 contract writers selected for the stamps. One of the 16 writers of Jinjiang was Smiling Cat, a writer of tanbi novels. What was unusual about this event was that while tanbi novels was (and still is) forbidden to be published in China,[2] tanbi writers were allowed to appear on the stamps together with writers of romances, swordsmen novels and so on. In fact, some tanbi writers, such as Tianlaizhiyi and Fengnong, were more popular than Smiling Cat, but their works contain more sex and violence than hers. And Jinjiang chose Smiling Cat, a less popular writer, as the representative of its writers for the stamps.

Here, we see the interesting and ambiguous situation of tanbi novels in China. On the on hand, tanbi is one of the several main styles of net novels in China. For example, in Jinjiang, a very large and well-known novel website, about a quarter of the most popular writes are tanbi writers. On the other hand, tanbi novels are forbidden to be published in China, so tanbi writers cannot share with other popular writers the benefit of the large book market. Tanbi culture, as a subculture,[3] like

[1] In July 2008, Snda Net Co. purchased three well-known novel websites including Jinjiang, Qidian and Hongxiutianxiang, and established Snda Net Novel Co. which possessed 80% market share of Chinese novel websites. In Dec. 2009, Snda Net Novel Co. purchased another famous novel website Rongshuxia and really became the monopoly in the market of Chinese novel websites.

[2] A few novels about ambiguous emotions between males were published sporadically in China. For example, *Bad Boys in Class 7 Grade 10*, wrote by Qixi, was published by Yunnan People's Publishing House in 2009. The novel is about the friendship and some ambiguous emotion between several boys. Some fojoshi regard it as a tanbi novel without sex description.

[3] In the 1940s, the sociologists of the Chicago School were the first to introduce the idea of subculture. Their research of subculture was about the culture popular in some particular groups in cites. Gay, rock music and hippies are typical subcultures. In the 19x0s, culture scholars at University of Birmingham studied the subculture of young

many other subcultures in China, shares the values of the mainstream culture and challenge it at the same time.

In this chapter, after the cultural background and descriptions of tanbi novels outside and inside of China are presented, the following questions of interest are addressed: why tanbi subculture is well accepted in China, what desires of young women are reflected in it, and how it challenges the mainstream culture.

15.2 History of *Tanbi* Subculture in Japan

Tanbi means "being addicted to beauty" in Japanese, and refers to aestheticism in the history of Japanese literature. Tanbi *literature*, also called "new romanticism literature," was a genre of literature in the beginning of the 20th century in Japan. It happened and developed on the basis of the objection to the influential naturalism literature at that time. The most famous tanbi writers were Nagai Kafu, Tanizaki Junichiro and so on. Tanbi writers argued that the ideas of naturalism writers such as "objectivism," "truth is more important than beauty" and so on would "oppress human beings' desire" so as to "lose beauty and human nature." Accordingly, they insisted on "acute mental and emotional sensibility" [Ye, 2009].

At the end of the 1960s, a new phenomenon happened to girl's comic. [4] Some comic writers, represented by Takimiya Keiko and Yamagishi Ryouko, began to draw comics about boy's love (Fig. 15.1). Due to its root in girl's comic, the boys in tanbi comics at that time looked like girls. Since then, boy's love comics have been popular among comic readers; most of them are young women and girls. They called these new comics *tanbi* partly because it had something in common with the tanbi literature at the beginning of the 20th century. And soon enough, tanbi novels appeared. Both the new tanbi

---

workers and pointed out that subculture, as a marginal culture, resisted the mainstream culture.

[4] Girl's comic, appearing in the 1950s, is the comic in girl's magazines. Most of their authors are women. Early girl's comic authors regarded girls from age 12 to18 as their readers. Then, the intended readership was expanded to girls from age 6 to 12 and young women from age 18 to 25. Most of girl's comics are about romance.

comics/novels and the old, traditional tanbi literature have much detailed description of mind, sense and body; appreciate pure love beyond ethics and social codes; and made use of sad and blue atmosphere. The women readers called themselves *fujoshi* which means "women indulged in something."

Fig. 15.1. Covers of the first *tanbi* comic, *The Poem of Wind and Tree*, Vols. 1-4 (from right to left, top to bottom) by Takimiya Keiko.

Under the influence of tanbi comics, many stories of boy's love appeared in *doujin*[5] comics and novels since the 1960s. Authors of doujin created many stories about love between two male characters in comics or video games, even though these characters did not fall in love at all in the original materials from which the doujin works were derived. As more and more stories about boy's love appeared in doujin comics and novels, the term *doujin* referred specifically to recreated stories about love between two male characters. In this sense, doujin is equivalent to "slash"[6] in the West. Fans of doujin are called *doujin's onna*. Here, we should pay attention to the fact that with rare exceptions, both writers and readers of tanbi are girls and young women, indicating that it is not male's desire but female's that is reflected in tanbi.

Those tanbi and doujin novels and comics which have too much description of sex are called *yaoi*. Most of tanbi and doujin texts in comics and novels have sex descriptions. However, for fujoshi, a moving and romantic love story is as important as the sex description, or even more important than it.

15.3 Two Types of *Tanbi* Stories and the Binary Opposition of *Seme* and *Uke*

To understand the historical and cultural background of tanbi subculture, it is necessary for us to know something about the description of gay in the history of Japanese literature. Generally, such gay stories can be divided into two types. One type is gay *gigolos* who behave passively and have to depend on their masters as women have to depend on men in an ancient society. Their masters are the single and total meaning and

---

[5] *Doujin* means independent creation in Japanese, and refer to "fan fiction," i.e., fictions created by fans of something. Authors of doujin use characters in comics and video games to create new stories, usually love stories. In the beginning, doujin was just stories about love between male and female. Afterward, more and more stories about boy's love appeared in doujin comics and novels.

[6] For example, the novels *The Persian Boy*, *Fire from Heaven*, and *Funeral Games* by Mary Renault and *24/7* by Xanthe are "slashes." Renault's three novels are about the love between Emperor Alexander and Hephaistion, his general. *24/7* is about the love between Fox Moulder and Walter Skinner, two characters in *X Files*, a popular TV program in the United States.

goal of their lives. Stories about noblemen and monks and their gay gigolos are recorded in many Japanese historical documents. For example, Tokugawa Iemitsu, the grandson of Tokugawa Ieyasu, had more than 100 gay gigolos.

Another type is young beautiful warriors who have wonderful swordsmanship. Different from gay gigolos, they are totally independent in their behavior and thinking, so their social position is not like that of women. Young beautiful warriors as men's object of desire are also recorded in many Japanese historical documents. The most famous example is the love between Maeda Toshitsune and Horio Tadaharu. In his historical novel, Ihara Saikaku[7] described the first date of the two warriors in detail.

Fujoshi divided the two male characters in tanbi stories into *seme* and *uke*. Seme means attack and is called the active side in the sex between two males; uke means accept and is called the passive side. By means of this division, fujoshi established the identities of the two males. Here, we see the binary opposition of seme and uke. As is known, binary opposition means ideology. *The binary opposition of seme and uke is the ideology of tanbi stories.*

The two types of gay stories are also the two types of tanbi stories. According to the position of uke, tanbi stories are further divided by fujoshi into two types: weak uke and strong uke. Weak uke, like gay gigolos, usually has lower social position than seme and depends on seme economically and emotionally. Differently, strong uke is usually on equal level with seme and has independent thinking and personal living goal. He does not depend on seme economically or emotionally. The division of strong uke and weak uke is the most basic division of tanbi stories. Presently in China, tanbi novels of strong uke are much more popular than those of weak uke.

---

[7] Ihara Saikaku (1642-1693) is a famous Japanese writer in the Edo Age. The subject of his novels is sex and love.

15.4 Texts about Gays as a Global Fashion

In the last dozens of years, political, economical and legal rights of homosexuals have been better approved and protected in many countries. Generally speaking, gays have been facing less and less discrimination in their careers. In many places in the world, laws have been passed to recognize gay marriage as legal. The global movement of homosexual resistance and the legalization battle is fading into history, or at least not as rigorous as before. However, and perhaps with regret from a literary point of view, the successful legalization of homosexuals also means that the movement of resistance upon homosexuality, as one of the many left-wing movements in the 1960s, becomes less and less rebellious. As commonly known, marriage is the outcome of our world based on patriarchy and heterosexuality. The legalization of gay marriage turned the homosexuals into good members of the mainstream society and stop challenging it.

It was in the background that texts about gays first came into the mainstream culture and eventually became a global fashion. In order to delete the challenge from the gay subculture, the mainstream culture adopted a strategy of removing the difference between homosexual and heterosexual and making homosexual look like heterosexual. Early texts about gays such as *In a Year of 13 Moons* (1978) and *Kiss of the Spider Woman* (1985) are part of the cultural criticism beginning in the 1960s. They had much discussion on body and gender. Nevertheless, texts about gays in the past decade have been becoming more and more like romance. In *Brokeback Mountain* (2005), *Milk* (2009) and *A Single Man* (2009), the logic is that we should not discriminate against gays because like us, they have moving love. The discussion on body and gender which indicate the identity of gays disappeared, and love, the subject of romance, blurred the identity of gays. Such texts were no longer the rebellious subculture of gays but just texts about gays. On behalf of romance, the female in it as the object of males was replaced by a male. The colonization of gay texts by mainstream culture created a new romance which satisfied the "new young women" in China. We will discuss about it in Sections 15.5 and 15.6.

As pointed out by Dai Jin-Hua [1999], after the Cold War, the globalization of the mass culture of the United States made the voice of cultural criticism becoming smaller and smaller, and some rebellious subcultures in the 1960s and the 1970s (such as Punk and Rock) are just fashions today; they are no longer challenging. The gay subculture is such a fashion. In the last decade, in spite of some criticism, films about gays were so popular that the Film Festivals of Berlin, Venice and Cannes all set up special award for queer films. Meanwhile, ambiguous emotions between males began to appear in mainstream films, too. A case in point is *Sherlock Holmes* (2009). As well as in many other countries, China is actively sharing the mass culture of the United States. Films and TV programs with ambiguous emotions between males are popular in China, with *Nishuihan* (2004), *Qiuxi* (2009), and *Sound of Wind* (2009) as examples.

In Japan, similar change even happened to gay adult videos, unlike formerly that there were just texts for gays. In traditional Japanese gay adult videos whose audience is mainly gays, the characters usually are a beautiful boy and an ugly middle-aged man. But in 2005, Coat West, the subsidy of Coat[8] in the West, tried a new type. In these videos, two beautiful boys (Fig. 15.2) took the place of a boy and a middle-aged man in the old type. They were immediately welcomed by the market, especially the market of fojoshi. Sho, Nagi and Hikaru, the three most well-known actors of Coat West, suddenly became the idols of fojoshi. Since then, more and more of such new videos have been produced. They are even called "gay adult videos for fujoshi." Of course, "gay adult videos for fujoshi" are not really just for fujoshi because gays like them, too.

More importantly, gay adult videos for fujoshi have another characteristic: sex with love. Sho, Nagi and Hikaru together played roles in two dozen videos.[9] After Hikaru quitted the job, Sho and Nagi continued costarring for several months. Fixed partnership made them look like lovers instead of just actors. These new videos are often shot in beautiful spots such as tideland, hot spring, skiing park and so on.

[8] Coat is one of the two largest companies producing adult videos; the other one is KO.

[9] The most well-known videos among them are *Luxe* (6 parts) and *Nagi and Sho* (3 parts).

Different from traditional gay videos in which sex occupies most of the video time, 2/3 time of these new videos are spent on showing the travels and the intimacy between the actors. The characters play joyfully and intimately, just like ordinary lovers. Moreover, through its official blog, Coat suggested to the audience that Sho and Nagi are lovers in their real lives.

Fig. 15.2. Sho and Nagi, two well-known actors of Coat West.

15.5 New Young Women in China and Their Desire

In their book *Slightly out of Character: Shonen Epics, Doujinshi and Japanese Concepts of Masculinity*, Megan Harrell and Mary College [2006] analyze the reason for the popularization of the tanbi comics in the 1960s in Japan. They point out that in the 1960s, Japanese women were changing from traditional, dependent women to modern women, but Japanese men whose typical image in mass culture was the salary man remained the same. These modern Japanese women felt unsatisfied with such men and began to construct new men mentally.

In the last decade, Chinese women were changing enormously. Unlike Japanese women, Chinese women have been economically independent since The People Republic of China was founded in 1949.

The government granted equal political and economical rights to men and women. In the last decade, many Western films and TV programs about love and sex of women were very popular among Chinese women living in big cities. A good case is *Sex and the City*. Meanwhile, some Chinese feminist scholars, Li Yin-He [10] for example, introduced Western feminist theories to China and discussed women's pleasure in sex.

According to Jacques Lacan,[11] the "phallus" is a symbol of social rights. As mentioned above, Chinese women have been sharing equal social rights with men since 1949. When they were after the pleasure in sex, they tried to share the bodily pleasure produced by the phallus. It is the reason why gender-neutralization has been a fashion in young women in big cities since 2000. In 2005, because of her gender-neutral image, Li Yu-Chun, the champion of a TV singing competition, became the most popular pop star among Chinese young women.

Undoubtedly, most of the independent young women in China are heterosexual. Thus, their desire of sharing the pleasure of the phallus cannot be satisfied in romances since in an ordinary romance, the object of phallus is the female herself.

As mentioned above, in the recent decade, no longer rebellious, texts about gays became part of the mainstream culture and were seen as chic. Meanwhile, Japanese comics have immense influence among Chinese teenagers. Hiding in the background young Chinese women began to write texts about gays as their new romance. According to interviews with some fujoshi, Chen Jia-Qi and Song Hui [2009] said in their book *Survey Report on the Influence of Japanese Comics*, "Women feel themselves at a disadvantage in sex, so they want to be a male and be loved by a male at the same time. …In order to make sex fair, they create sex between two males."

In this new romance, young women are free to regard the male as the object of their desire and enjoy the pleasure produced by the phallus. An interesting case is that several popular Chinese female tanbi writers,

[10] Li Yin-He is one of the most famous feminist sociologists in China. Her book *Love and Sex of Chinese Women* [1998] is very influential among women students in universities.

[11] Jacques Lacan, the famous French poststructural theorist, argued that the social right is represented by "phallus" which is the center of the society. In this sense, females can share "phallus" and males may not have "phallus."

e.g., Tianlaizhiyi and Jianzoupianfeng, falsely claimed they were men. Maybe they hope to be men. However, more importantly, readers hope to see male tanbi writers. And male writers are easier to be popular than female writers. Fujoshi might experience more real pleasure produced by phallus in novels by male writers than they do in novels by female writers.

In fact, fujoshi show some unconscious resistance to romance and the mainstream culture based on heterosexual. For example, many fujoshi said, "I feel great dislike when I see a boy and a girl fall in love." For this reason, they try their best to draw a clear dividing line between tanbi novels and romance. They argue that "We do not want weak uke!" and "Whoever watches gay adult video is real fujoshi." As mentioned in Section 15.3, the position of weak uke is like that of the female. According to fujoshi's concept, weak uke is female and tanbi stories about weak uke are romance. However, factually tanbi stories about strong uke are romance too. (See Section 15.6 for details.) For fujoshi, gay video means real male body. As we know, romance is the binary opposition of the subject and the object based on love. It is unimportant whether the binary opposition is between male and female, or between seme and uke. Tanbi novels are thus a new romance in which the binary opposition is between seme and uke. In this sense, for fujoshi, gay video is of course romance. Here, we can draw the conclusion that fujoshi's discrimination of tanbi from romance is an attempt without meaningful result.

15.6 Fatherly Love in the New Romance

In *Powers of Desire: The Politics of Sexuality*, Ann Bar Snitow et al [1983] pointed out that in romance, a self-closed structure, except love, nothing is important. The only function of the stereotyped stories is to offer a place where the female will wait for the love of the male. This is also the characteristic of tanbi novels partly because tanbi comics were developed from girl's comics in the 1960s in Japan (see Section 15.2).

More importantly, tanbi novels and romance share the same love—love to father and love from father. As mentioned above, phallus means social right, so females can share phallus and males may not have phallus.

In this sense, father is the male who have phallus, and he is the real governor of the society. In other words, the society is patriarchal rather than male-dominant. Romance is just the outcome of the patriarchal heterosexual mainstream culture.

Tanbi novels, as a romance, necessarily share the subject of fatherly love. Since his position is like that of female, the weak uke's only meaning is to be the object of the male's desire. Of course, the male's desire is the desire of fujoshi who try to share the pleasure produced by phallus. Even fujoshi are aware that stories about weak uke are romance, so they try to abandon weak uke. However, the so-called strong uke, independent in behavior and emotion, does not look like female. This is the reason why fujoshi insist that stories about strong uke are not romance, but are real tanbi.

However, in the final analysis, these stories about strong uke are romance too because they are still the binary opposition based on fatherly love. In fact, the independence of strong uke is similar to that possessed by modern women. It is just based on the independence of themselves that fujoshi construct the independent strong uke. Unlike weak uke, stong uke is not female, so he can be free to love father and be loved by father.

However, fujoshi show confusion on fatherly love. On the one hand, they are deeply attracted by the fatherly love constructed by romance. On the other hand, unconsciously feeling dislike to the patriarchal heterosexual mainstream culture, they cannot justify female's desire to love father and be loved by father. In order to delete the confusion, they construct in texts an independent male who loves father and is loved by father. The male is the strong uke that satisfies fujoshi's desire to fatherly love. The desire of fujoshi has something in common with the fact that tanbi novels about love between father and son are very popular in China.[12] Of course, as romance, tanbi novels do not really care about ethics. However, more importantly, it is the real desire of fujoshi. Here,

[12] When tanbi novels about love between father and son began to appear in China, fujoshi had bitter arguments on it. Many fujoshi showed intolerance of incest. However, afterward, the opposition made no difference to the popularization of such novels.

the binary opposition between strong seme and strong uke is the binary opposition between father and son.

In the modern discourse, the binary opposition between father and son is not the same as that between male and female because son will become father sooner or later, but female can never become male. In this sense, father and son are equals, and they are real males. It is for this reason that fujoshi do not think stories about love between father and son are romance.

However, stories about love between father and son are romance, too, because in such novels, the son does not become father. In fact, in tanbi novels, uke does not become seme, and the son never share the pleasure produced by phallus. The son is a male just because he will become father; if he never becomes father, he is not son but female. Fujoshi try to construct a real male as uke in tanbi novels, but the fact that tanbi novels are romance decides it to be a mission impossible.

15.7 *Tanbi* Novels in China: Writing and Publishing

Under the influence of Japanese tanbi subculture, at the end of the 1990s, some Chinese girls and young women, mostly from age 15 to 25, began to write tanbi novels on websites. They are the earliest fojoshi in China. However, even now, tanbi novels have been forbidden by the Chinese government to be published because the gay subculture is still regarded as pornographic in China, and pornographic works such as films and novels are illegal. By the same reasoning, romances with much description of sex are allowed being published since they are not considered as pornographic novels, but erotic novels. As mentioned above, a few isolated novels about ambiguous emotions between males were published in China. But these are too small in number to be taken into account.

In 2003 and 2004, several novel websites brought economic interest to writers of tanbi novels whose works could not be published officially in China. They are Jinjiang, set up in 2003; Qidian, in 2003; and Myfreshnet, based in the United States and paying attention to the market of China in 2004. Every writer was free to publish his/her works on these websites. Editors of the websites chose some popular novels from abundant works and signed agreements with their writers. Then, the selected writers became so-called VIP writers and could gain payment from the websites according to the number of words they published because readers had to pay to read VIP novels. In fact, earlier than that, some novel websites such as Rongshuxia, set up in 1997, and Hongxiutianxiang, set up in 1999, began to pay VIP writers at the end of the 1990s. However, they did not publish tanbi novels. So writers of tanbi novels had to wait until 2003 and 2004 to receive any payment.

Right following the VIP payments, with the help of the websites, Chinese tanbi writers began to publish books in Taiwan. In 2004, Ueishiang Culture Press in Taiwan published *The Slave* by Fengnong, a very famous tanbi writer. It was the first Chinese tanbi novel ever published in book form. Afterward, following Fengnong many Chinese tanbi writers published their novels in Taiwan. Their income increased by a large margin for this reason.

Apart from publishing books in Taiwan and signing agreements with websites, tanbi writers made money by printing their novels and selling them to readers directly or through shopping websites. These are so-called "personal books" (Fig. 15.3). In China, the ISBN numbers for books are rigorously controlled by the government. The "personal books" do not carry ISBN numbers and are not officially "published." In China where citizens are not free to publish books, the "personal books" run oppose the government's control of publishing.

The economic factor did help the popularization of tanbi novels since 2005. In spite of that, it is necessary to mention that even now, the payments due the tanbi writers are very low. All of the three main means of income together do not make abundant economic interest. The Taiwan publishers seldom print a tanbi novel with more than 1,000 copies so that the royalty for the writers is very limited. In addition, the Chinese government often cleans up the net sellers who sell tanbi novels. Pirated books seriously influence the quantity of sale of tanbi novels published and printed by writers themselves. Many VIP novels are illegally copied and published on other free websites, reducing the number of readers who pay for the VIP novels and thus the payments received by the VIP writers. Consequently, the income of tanbi writers is much lower than that of other popular writers whose works can otherwise be published in China.

Fig. 15.3. Covers of *The Dawn*, a two-volume "personal book" by Zhouerfushi.

Tanbi novels cannot share the large book market of best sellers in China. This is the reason why tanbi writers are usually young, aged from 15 to 25. Writing tanbi novels, unlike writing other best sellers, cannot be a useful means of making a living. Only girls and young women without heavy living pressure can write tanbi novels in their spare time. For this reason alone, the tanbi subculture has not developed a whole system of production and consumption like that in a mass culture. It shows the opposition tanbi experiences as a subculture even though fujoshi are unconscious of it; they may have no other choice even if they are aware of it.

15.8 Conclusion

In the 1960s, developing from girl's comics, tanbi comics appeared in Japan. In other words, tanbi novels originated from romance. Since the end of the Cold War, the gay subculture has gradually entered into the patriarchal heterosexual mainstream culture and been constructed as a global fashion. Texts about gays were no longer a cultural criticism as before and became a new romance—the binary opposition based on love. In the last decade, sharing social rights with males, young Chinese women began their desire to share the pleasure produced by phallus. It was through the back door that Japanese tanbi comics and novels slipped into China and became more and more popular among young Chinese women.

The end of the Cold War implies the end of the left cultural criticism beginning in the 1960s. Many subcultures which had been rebellious in the 1960s gradually became part of the mainstream culture. In this environment, the tanbi subculture is just a new romance in the Chinese mainstream culture. In this new romance, fujoshi regard the male body as the object of their desire and experience the pleasure produced by the phallus. In addition, in this new "game," they can be free to love father and be loved by father—without any confusion resulting from their unconscious dislike of the patriarchal heterosexual mainstream culture.

However, even though fujoshi are unconscious of it and may have no choice, tanbi, as a subculture, still shows some opposition to the mainstream culture. For example, the tanbi "personal books" do oppose the government's control and the commercialization of publishing.

References

Chen, Jia-Qi & Song, Hui [2009] *Survey Report on the Influence of Japanese Comics* (People's Publishing House, Beijing) p. 75.

Dai, Jin-Hua [1999] *Invisible Writing: Studies on Chinese Culture in the 1990s* (Jiangsu Renmin Press, Nanjing)

Harrell, M. & College, M. B. [2006] *slightly out of Character: Shonen Epics, Doujinshi and Japanese Concepts of Masculinity* (Women's Studies Quarterly, Feminist Press, NY)

Li, Yin-He [1998] *Love and Sex of Chinese Women* (Today China Press, Beijing).

Snitow, A. B., Stansell, C. & Thompson, S. [1983] *Powers of Desire: The Politics of Sexuality* (Monthly Review Press, New York).

Ye, Wei-Qu [2009] *History of Japanese Novels* (Peking University Press, Beijing) p. 281.

16

Objects in Art and Science

Nigel Sanitt

Up to the nineteenth century western art was dominated by the idea that art (paintings in particular) represented objects. This view termed *mimesis* assumed that (1) the world exists before and independent of the work of art; (2) the medium is a neutral means of representing objects. Mimesis today is only a small part of art theory, with representation of the world in art being seen as culturally and historically variable. The mimetic view of science (if I can refer to it as such) is that there is a world independent of the scientist and that science is neutral and objective. This is the dominant view of science today. In spite of enormous technological and theoretical progress in science during the last hundred years, it seems to me quite surprising that this mimetic view of science still holds sway and that in the sense I have described it, scientists do not seem to have caught up with the art theorists.

16.1 Introduction

This chapter is not about art theory (and I apologize to art theorists who think that I have inaccurately simplified their subject). My aim is to look at objects in science and use images (paintings and photographs) as a metaphor to highlight problems in interpreting scientific theories and the objects they claim to create.

As an alternative to the conventional view of scientific thinking, I describe a scheme which has as its starting point the idea that questioning is at the root of science (Problematology) and rejects an object-oriented mimetic view of science.

A good place to start talking about objects in science is with the familiar concept of "particles". It is also a good time because the Large

Hadron Collider (LHC) is operational at CERN, Geneva, and particle physicists are hoping that many new particle discoveries will be made in the near future.

Anyone can download from the internet a comprehensive list of known particles. The latest list[1] describes well over a hundred particles and no doubt in the coming years the LHC will add many more.

But what exactly are particles? If you ask a physicist this question, you will often be greeted with a somewhat vague answer, involving quantum states and fields, particle-waves and a lot of "arm-waving". But as David Mermin points out: "quantum states are calculation devices" and further: "quantum fields are useful mathematical tools" [Mermin, 2009]. Useful they may be, but are particles actual objects in the world or just useful fictions. Perhaps we can look to the world of art for inspiration.

16.2 Mimesis in Science

Let me start with what, at first, appears to be a straightforward example. In 1884 the American artist John Singer Sargeant, who was living in Paris, France, painted a portrait entitled *Madame X* (Fig. 16.1). In fact, the subject of the portrait was Madame Pierre Goutreau, a wealthy socialite and hostess, who in today's terminology would be classed as an "A–list celebrity". The "original" painting can be seen at the Metropolitan Museum of Art, New York, USA. One might say that this is a clear example of mimesis: the "object" of the painting being the woman depicted.

Note, however, that I put the word "original" in inverted commas. In fact, this is not how the original painting appeared. The right shoulder strap of the lady's dress was missing in the original painting. This caused an outrage when the painting was first exhibited. It was claimed at the time that Madame Goutreau (a highly respected and respectable woman!), was made to look like a prostitute, and the artist was forced to remove the painting from the exhibition. It was only allowed back into the

[1] Downloaded October 2009. Detailed properties of 124 known particles and 39 hypothetical particles were listed.

exhibition after Sargeant had painted in the missing strap, which is the version of the painting we see today.

Fig. 16.1. John Singer Sargeant, *Madame X*, 1884 (Metropolitan Museum of Art, New York).

To the Parisians of the time, the original painting depicted not simply the woman, Madame Goutreau, but a depraved and morally degenerate image, which grossly offended their sensibilities. Today, most of us cannot see what all the fuss was about. But clearly the passage of time and the cultural and moral distance between us and the Parisians of the 1880s, means that there is no such thing as straightforward mimesis.

An example from astronomy: The National Aeronautics and Space Administration in the USA (NASA) produces many images of celestial objects none more exquisite and majestic than that of the Whirlpool Galaxy (M51), shown in Fig. 16.2, left. If ever there was an example of straightforward mimesis in science, this image represents a clear candidate; M51 is out there, and this is its image.

Fig. 16.2. *Left*: Whirlpool Galaxy (M51), NASA. *Right*: Whirlpool Galaxy (M51), William Parsons (Lord Rosse), 1845.

Alongside this image is another of M51 (Fig. 16.2, right), which is a drawing executed by William Parsons, the Third Earl of Rosse, in 1845. Lord Rosse's 72-inch telescope was the largest in the world at the time. He was a gifted observer and artist and the drawing was the result of a number of night's careful observations. The point about these two somewhat different images of M51 is that if you look through a telescope at M51, then what you actually see is much closer to Lord Rosse's image than NASA's—Rosse drew what he saw.

The problem is the human eye. M51, in common with spiral galaxies, has a central core which is much brighter than the surrounding spiral arm structure, and the human eye finds it very difficult to accommodate large differences in brightness over a single image. Even photographic paper has its limitations and in the NASA image of M51, the central core is overexposed in order to show the very wispy spiral structure.

If that were all then, apart from a brightness imbalance, we could say that the NASA image corresponded to an object, which is depicted. However, consider Fig. 16.3, which is a photomontage of a floating iceberg. How much of an iceberg which is hidden from view under the sea depends on the salinity of the seawater, but on average is about 90%.

So when you see an iceberg floating in the sea you are only looking at about 10% of what there is.

Fig. 16.3. Photomontage of what a whole iceberg might look like. [Wikimedia Commons.]

Surprisingly, approximately the same is true of galaxies. Generally accepted by cosmologists is the idea that all the matter in the universe we can detect, not only at optical wavelengths but also at other wavelengths, corresponds to only about 7% of all matter. Furthermore, nobody knows what the other 93% of matter is made of—maybe the LHC at CERN might answer this question.

In spite of its great beauty and a prime example of *AstroArt*, as mimesis, the NASA image of M51 is only a very rough and ready taste of what is actually out there. Is the referent M51 just a product of language, or is there an ocean of icebergs out there, of which M51 is the label we give to one of the shards we see sticking up out of the water?

16.3 Time and Language

The problem with envisaging science as the creation of "objects" is that a universe full of objects is rather a boring place. Nothing much happens unless the objects actually do something. In a roundabout sort of way, what I am saying is that the element of time or process in such a world seems to be either missing or of reduced importance. We can ask: how is time or action represented in art and science? Fig. 16.4 shows frozen moments of time.

Fig. 16.4. *Left*: Tesla Coil Sparks. *Right*: Auguste Rodin, *The Kiss*, 1886 (Rodin Museum, Paris). [Both Wikimedia Commons.]

In normal dry air, the electrical resistance of the air determines the voltage required to produce a spark of lightning.[2] Fig. 16.4, left, is a photograph of a Tesla coil at the very instant that a spark is produced. The frozen moment is captured in the image just as in Fig. 16.4, right; Rodin captures the intimate moment of *The Kiss* in his famous sculpture. There are different kinds of "sparks" depicted in the two images, but the representations in both cases are not the objects involved—the Tesla Coil

[2] At standard temperature and pressure 3.3 million volts per meter.

or the lovers—but the actions: time is of essence; dynamical relationships are important, not objects.

Interestingly, the images of M51 (Fig. 16.2) can also be interpreted as depicting “process”. The spiral arm features captured in both images are, in fact, the captured “instant” of a stellar traffic jam. The arms are “density waves”. These are similar to density waves that describe the bunching of vehicles on a crowded motorway. In a spiral galaxy the waves of compression of the star-forming gas result in “waves” of stars being produced, which gives rise to the spiral arm structure. So, in this sense, even the “portrait” of M51 shows a frozen moment of time.

The idea that language is somehow a mirror on nature with names standing for real objects in the world, which are somehow “out there” waiting to be discovered, is false. Meaning is “continuously created” in an open-ended process of relationships and the networks of these relationships can be mapped.

Rather than thinking in terms of objects, we can think in terms of networks of questions and answers. Meaning thus arises out of scientists integrating[3] their answers and questions within the networks they create. There are, in a sense, no objects, but only relationships between questions, mediated by answers.

16.4 Science Questions

I want now to describe a model for scientific theorizing that takes as its basis the idea that questioning is at the root of science and therefore merits study.[4] To be more explicit, I refer to a certain kind of questioning —the open-ended variety. Questioning of the type: “What is one plus one?” or “What is the capital of England?” are closed questions. These questions have definite answers: “two” and “London”. They do not, of their nature, lead to other questions. Their answers solve the problem generated by the original question. These kinds of questions occur in science, but only as part of a mathematical or deductive process. A

[3] I refer here to the theory of meaning termed “integrationism”, initiated by Roy Harris, on which these ideas on meaning in science are based [Harris, 2005].

[4] For questioning as the fundamental reality of the human mind—Problematology—see the works of Michel Meyer [1979; 1995; 2007].

mathematical equation in science is a shorthand way of introducing a logical connection. It answers the question (or questions) of how one thing is related to another in a purely systematic, logical or deductive way. It is a bridge across a deductive gap.

The other type of question—and the one I want to concentrate on—is of the open-ended variety. Examples are: "What is an electron?" "Why is it raining?" "How did the Earth form?" These questions lead to other questions and they do not occur in isolation—they are always part of a sequence of questions. These kinds of questions are the bedrock of science.

The question is: how scientific questions fit together with the answers that our experiments and observations come up with? We need to look at the relationship between answers and questions or as I suggest answer/question pairs. The phrase "answer/question pair" has a peculiar kind of dissonance. The reason is that the words "answer" and "question" are in the reverse order that they normally occur in speech. We may often refer to questions and answers, but we do not usually, if at all, refer to them in the reverse order.

So what is an "answer/question pair" and why is it important in looking at the way science is structured? It does seem perverse to imply that the answer comes first in reply to the question: What comes first, the question or the answer? However, just as in the proverbial chicken and egg problem, we can say that a chicken is the way an egg creates another egg, I would say that an answer is the way one question leads to another question. An answer to an open-ended question always leads to another open-ended question. Thus in the chain of reasoning, why not treat each answer and its subsequent question as the unit of importance. In this way questions and their subsequent answers are replaced by questions and their preceding answers—we, as it were, face the other way.

So what difference does this way of looking at questions and answers make? Instead of looking at the answer to a question as of prime importance, we look behind every question, to see how it arose. The effect of this change of perspective is that questions are regarded as more important units in the chain than answers. This stability or invariance of questions, not shared by answers, is, I believe, foundational to science.

In addition to the distinction between open-ended and closed questions, concentrating on the former, I now introduce within open-ended questions a further distinction: In a chain of questions and answers, I describe the structure in terms of "answer/question pairs". Described in this way, it follows that any finite chain of such units must start on an answer and end on a question. If we look at the end of the chain, what does it mean to end on a question?

In science this represents the situation that we do not have answers to all questions. Any theory, however well accepted, will have unanswered questions. The answers (and future questions) to these questions await the discovery of further theories. Just as the questions are open-ended, so are the theories—we are on an infinite staircase. Not only is it reasonable that we should have open-ended questions in science, but science would not be science without them.

The problem with the scenario I have outlined, however, is not the observation that chains end on questions, but the problem of how a chain can start on an answer (or an "answer/question"). After all, it is deeply ingrained in our ways of thinking that you can only start a chain of questions and answers on a question. It would thus seem at first that the idea of looking at questions and their preceding answers collapses when faced with a "starter" question which appears devoid of a preceding answer.

The way out of this conundrum is to realize that such generative questions or "empirical questions" as I call them, are ones where their preceding answers (and questions) are suppressed. This suppression or "black-boxing", as I define it, refers to science which is accepted or taken for granted. The preceding questions and answers have become presuppositions. Thus all theories start with presuppositions and empirical questions, as defined.

So why theories? Why science? Why presuppositions? These questions are not usually addressed in philosophy of science, because they are grounded in the human psyche. Maybe so, but we are part of nature and we have evolved to ask questions. The answer to what is behind question-asking is to refer to the somewhat question-begging term "curiosity".

So what exactly is curiosity? Purely in terms of appearances, we can ask: How do entities who are exhibiting curiosity behave? The answer to this question would be something like: curiosity is manifested by some type of exploratory behavior, where an entity initiates an interaction with its environment. The difficulty with this definition is that it disguises the fact that it is us—the observer—that is putting meaning into the world, in the way we view other entities—describing them as exhibiting curiosity. An alternative definition of curiosity would be as a drive—a biologically derived compulsion. As eating and drinking satisfy the drives of hunger and thirst, the question then becomes: is curiosity a biological drive which we need to satisfy? One can see that hunger and thirst are important drives, as if we did not eat or drink we would die: but why curiosity? So that we can do science! I do not think that that is a satisfactory answer. Clearly, in common with the rest of the animal kingdom, we are hard-wired for curiosity. In the end at some deep level, we have to just accept that we are a curious species—we ask questions and that part of this process is the formal backdrop to science.

So if science is a collection of questions, or as I have described it, a collection of "answer/questions", how does the whole thing fit together? What are the rules, if any, and how does this relate to science in the laboratories and Institutes as we know it? Before embarking on these questions, I want to look at the question of meaning in science, under the terms "object" and "process".

Quite a lot has been written on the subject of meaning in general and meaning in science, in particular. I want to concentrate on one aspect of meaning in science: namely the object/process division.

"Process" implies "time" by its very nature. "Object" implies stasis —things which do not vary over time or are in some sense invariant. The division, perhaps, is not so useful for our present purposes. In science, objects seem to be terribly important. Many scientists are used to wondering about neutrons or neutrinos, stars and galaxies to the point where it seems that science is dominated by the discovery and classification of objects. Even if you look at processes, they are usually directed in terms of how objects behave in time. Meaning, on the other hand, does not come from the identification of objects as such, but in terms of relationships and interactions.

16.5 Science and Truth

Truth is a difficult concept to pin down. Attempts in the sciences to define "scientific truth" always crash in flames. It is impossible to avoid, at some point in any discussion, "begging the question", when it comes to defining scientific truths. The reason is that there is no "truth" in science. By this, I mean that any proposition about the way the natural world will behave will always be open-ended, and therefore subject to change. Tautological statements, which cover mathematics, are not statements about the way the world will behave, and so may have truth values; they are unlike statements about the physical world, which cannot have truth values. By excising truth from the structure of science in this way, I avoid all the nasty (or I would say impossible) problems about scientific truth.

One of the really useful characteristics of questions is that they have no truth values. You cannot say about a question: Is it true or false? My point here is that in dealing with networks of questions, we do not have to worry that we are going to generate any scientific truths, because the questions themselves have no truth values.

But what about the answers and any mathematical parts to scientific theories or explanations, which may be hiding among the network? Answers certainly have truth values and as I have already stated, so do mathematical statements. Herein lies the truth—literally. Scientific theories thus—according to this scenario—can be expressed in terms of networks of questions and answers: the basic elements being the questions, with the answers forming (temporary) links between questions. Presuppositions and accepted theories, on which new theories are grounded, are hidden, suppressed or black-boxed within generative or empirical questions—as I refer to them. The answers which link the system and which include factual and mathematical statements reside in the network on a temporary basis. The system is always open to reformulation, reinterpretation or capitulation to other theories which may come along. The interesting thing is that even though answers may blow away in the wind, the questions have a much longer shelf life—in fact they never go away. They may not be relevant any more to a particular theory or may fail to be incorporated into any new structure, but they have a characteristic quality of being invariant and unchanging. It is thus their very lack of having a truth value that is their strength and which guarantees their longevity.

Many scientists would argue that they are not really bothered about the "truth" of a theory. A theory, for them, should encompass sufficient experimental results and generate enough predictions to be useful. As far as the irrelevance of "truth" is concerned, one can get into semantic somersaults in terms of the "validity", "truthfulness" or "approach to truth" of a theory and which of these words substitute for the "real thing". Once it is accepted that there is some ideal that is unattainable, at least for now, then a reasonable response is to turn the quest for the ideal as the goal in itself. Thus Popper's model [Popper, 1972] of theories converging towards, but never actually arriving at the truth, is probably accepted by most scientists, who at the same time have at the back of their minds the ultimate goal of a final "theory of everything".[5] In the meantime, they plug the numbers into their theories, generate predictions, which sometimes work and sometimes don't, and march towards the light. To be fair, science has marched a long way, but there will come a time when scientists will have to move from the adolescent to the adult stage.

16.6 Science Networks

So what are the "rules of the game"? If the structure or structures of science are represented by networks of questions, then what are the structural rules? How do we know if we have a valid structure? And by introducing the concept of "valid structure", have we not introduced the idea of truth in science—just what we wanted to avoid!

I take the last point first. I have made it clear that truth does reside in the system I have been describing, but I have taken care to show that it is only in certain areas that this is the case, while in other areas it is not the case. Thus within scientific answers and mathematical arguments, true (or false) statements abound. In addition to this I would add that when describing the structure of a network—as a mathematical structure

[5] See Chap. 20 in [Gardner, 1985].

governed by rules—the associated mathematical rules correspond to a system of truth values. But this does not impinge at the lower level where one looks at the questions themselves that may make up the system—they do not have truth values.

So what are the rules? There are three:

1. The first is the generative rule that science starts with empirical questions. Our presuppositions and existing meaning structures come together into a system which scientists start to weave into theories.
2. The second or open-ended rule is that any chain of scientific questions and answers cannot produce a tautology. Yes, mathematics is itself tautologous, but mathematical aspects are black-boxed within the system, and the open-ended questions cannot shut off further questions. Thus in building up a scientific structure of questions and answers, we do not have any closed loops or cycles of question and answer chains.
3. Finally, there is the gain rule. Scientists produce theories and science progresses by trying to incorporate more of the world within our meaning and understanding. Any theory which fails to do that will not be considered worthwhile or even scientific. In terms of the networks of questions and answers I have been describing, I interpret this rule to mean that the number of empirical questions that a theory incorporates must be at least greater than the number of new questions that the theory has thrown up.

These three rules and the structural model I have been describing for science, can be easily represented mathematically—the relevant area of mathematics being "graph theory".[6] The three rules have analogous descriptions in graph theory, which means the science structure, as I have been describing it, can be translated into a straightforward mathematical model.[7]

---

[6] A branch of Mathematics founded by Leonhard Euler (1707-1783).
[7] It is not my intention in this chapter to go into mathematical detail; interested readers can refer to [Sanitt, 1996].

16.7 Science and Human Nature

And finally, what of beauty? In 2009 a mathematical journal[8] held a competition for the most beautiful equation, which was won by Euler's identity[9] (see Fig. 16.5). Some might say that compared to Botticelli's birth of Venus, for example, Euler's identity: $e^{i\pi} = -1$, cannot really be classed as beautiful, although the latter is more popular on the internet.[10]

Fig. 16.5. Euler Plant (Adnan Masood's Blog).

In the next few years I look forward to scientific discoveries that will be made and new theories that will come to the fore—particularly as a result of results that may come from the LHC. But if the outcome of all the effort is just to add more particles to the list of known particles, then science will have fallen short of expectations. What is more important is to think about things in a new way, not just new theories, but new kinds of theories.

---

[8] *Mathematical Intelligencer*, June 1, 2009.
[9] The same Euler who founded graph theory.
[10] October 2009: A comparison of hits on Google: Botticelli/Venus 526,000, Euler/identity 724,000.

The scenario I have been describing for science consists not just of scientific objects, which are useful tools, but of human curiosity driving the creation of networks of questions directed at the world—a world which includes ourselves.

References

Gardner, M. [1985] *The Whys of a Philosophical Scrivener* (OUP, Oxford).
Harris, R. [2005] *The Semantics of Science* (Continuum, London).
Mermin, N. D. [2009] "What's bad about this habit," *Phys. Today*, May, 8-9.
Meyer, M. [1979] *Découverte et Justification en Science* (Kllincksieck, Paris).
Meyer, M. [1995] *Of Problematology: Philosophy, Science and Language*, transl. Jamison, D. (University of Chicago Press, Chicago).
Meyer, M. [2007] "Problematology: A new paradigm for thought," *Revue Internationale de Philosophie* **61**(242), 349-352.
Popper, K. R. [1972] *Objective Knowledge* (Clarendon, Oxford) p. 195.
Sanitt, N. [1996] *Science as a Questioning Process* (IOP Publishing, Bristol).

17

Su Dong-Po's Bamboo and Paul Cézanne's Apple

Lui Lam and Li-Meng Qiu

Su Dong-Po (1037-1101) of the Song Dynasty is arguably the most well-known poet and writer in China. He is also a distinguished painter; he liked to paint bamboos and rocks. Unlike his contemporaries and painters before him, the leaves in Su's bamboo painting are not necessarily attached to the stem. Paul Cézanne (1839-1906), a French post-impressionist, is recognized by Picasso, Matisse and many others as the father of modern art. He went beyond impressionism and painted many things including apples. Both these two artists tried to go beyond the appearance and show the *essence* of the objects they painted, in their own new ways. It was not by accident that these two painters—one from the East and the other from the West, separated from each other by about 800 years—had the same idea about painting. There must be something basic behind this. As shown in this chapter, the mechanism behind their techniques is based on how we see things, the cognitive science of vision in the human brain. The fact that Su's style was not adopted as mainstream, unlike that of Cézanne, is discussed; it is related to the unique nature of China's ultra-stable feudal system in the past, in which science and technology are implicitly or explicitly discouraged. Finally, the possible origin of Dong-Po Pork, for which Su is also famous for, is presented in an appendix.

17.1 Introduction

Su Dong-Po (1037-1101) from China and Paul Cézanne (1839-1906) from France (Fig. 17.1) lived about 800 years apart from each other, in two different continents. Both are pioneering painters. While Su is primarily a poet and writer, Cézanne is shy in writing [Cézanne, 1995, p. 9].

Fig. 17.1. Su Dong-Po (left) and Paul Cézanne (right, c. 1861). Su designed the hat he was wearing.

The preferred painting subjects of these two artists are quite different. Su painted mainly bamboo and rock, while Cézanne did still life, landscape, portrait, and nude (which he did without live models, for unavailability or his shyness, or both).

Both of them were nature lovers. Yet, this is not the only thing the two had in common, in spite of their very different personal character and career (see Sections 17.2.1 and 17.3.1 below). In their paintings, they were not satisfied to reproduce the likeness of the objects they painted, but instead strived to capture the very essence of those painting objects. And they invented new ways to do that.

Technically, Su used ink brushes and his black-and-white paintings were always done with simple strokes. In his bamboo paintings, some of the leaves are detached from the stem, creating a style quite different from those of his contemporaries. For Cézanne, he used oil brushes. His paintings were done with swift strokes with "planes" of color, reducing the objects to their basic geometric forms as he visualized them. And in so doing, he ushered in the era of modern arts.

In the following, the life and career, art view and their innovative painting style are presented, respectively, in Sections 17.2 and 17.3. The neurological basis of their art is discussed in Section 17.4. In Section 17.5, the fate of these two artists' painting styles, with the social background, is contrasted and explained. Finally, since many Chinese learn about Su's name through Dong-Po Pork, if not his arts, our findings about the origin of this delicious Chinese dish—including the cooking recipe—is given in an appendix.

17.2 Su Dong-Po and His Bamboo

Su Dong-Po's career, art view and painting innovations are presented.

17.2.1 *Su's Life and Career*

Su Dong-Po,[1] a poet, writer, painter and calligrapher, lived in the Northern Song Dynasty (960-1127). His life and career are summarized in Table 17.1.

Table 17.1. Chronology of Su Dong-Po's life and career.[2]

Year	Age (yr)	Event
1037	0^+	Born Jan. 8 in Meishan, Sichuan Province
1054	17	Married first wife, Wang Fu, 14 years old
1056	19	Ranked no. 2 in *juren* national exam. in capital, Changan
1057	20	Ranked no. 2 in *jinshi* national exam. in Changan; mother died; returned home
1061	24	Got his first job as assistant magistrate in Fengxiang
1064	27	Met Wen Tong, a bamboo painter
1065	28	Became Secretary of Dept. of History in capital; wife Wang Fu died at age 25
1066	29	Father died; returned home
1068	31	Married second wife, Wang Run-Zhi, 19 years old; moved family to Changan; resumed Secretary of Dept. of History

[1] Su is the family name of Su Shi; his literary name is Su Dong-Po, which is spelled differently in old literature as Su Tungpo [Lin, 1947], Su Tung-p'o [Sirén, 1935; Watson, 1994] or Soo Doong-Bô (*The Columbia Encyclopedia,* 2004, Columbia University Press). A short account of his life is given in [Watson, 1994, pp. 3-12]; a long account, [Lin, 1947]. (Note that in Lin's book, he records lunar calendar dates in solar calendar terms.) A thin Chinese book on his life and arts, with many of his calligraphy reproduced and a very useful chronology included, is [Yang, 2006]; a thick Chinese book is [Li, 1996].

[2] Adapted from [Yang, 2006, pp. 110-121]. For the same year, the number listed in our Age column differs from that in Yang's book, since he follows the Chinese custom in age counting (for example, he lists Su as one year old in 1036). Similarly, in our Event column, we reduce the age number by two from that in Yang's book.

1071	34	Opposed emperor's reform movement; sent to Hangzhou as deputy magistrate and stayed there for 3 years
1074	37	Took in Wang Chaoyun, age 10, as servant; moved to Mizhou as magistrate
1077	40	Became Xuzhou magistrate
1079	42	Became Huzhou magistrate; in prison (about 6 months) for writing poems and essays deemed politically incorrect
1080	43	Banished to Huangzhou, with a job but no pay;[1] took in Wang Chao-Yun as concubine
1084	47	Assigned to Ruzhou; visited Lu Mountain in transit
1085	48	Assigned to Changzhou, then Dengzhou, finally back to Changan as Secretary to Premier's office
1086	49	Became Secretary to Emperor (*hanlin*) for 3 years
1089	52	Asked and got transferred to Hangzhou (due to political conflicts in office)
1091	54	Back to Changan as Minister of civil service (for 3 months); asked and got transferred to Yingzhou
1092	55	Transferred to Yangzhou; returned to Changan as Minister of war and Minister of education; Empress, his political support, died
1093	56	Second wife Wang Run-Zhi died at age 44; transferred to Dingzhou as Commander
1094	57	Exiled to Huizhou, Guangdong Province (his old writings in his *hanlin* days were judged politically incorrect), bringing concubine Wang Chao-Yun (became third wife later) and third son along;[2] started building an elaborate mansion the next year
1096	59	Third wife Wang Chao-Yun died at age 32
1097	60	Mansion completed; banished to Qiongzhou, Hainan Island (the most southern part of China) when the emperor thought Su was too happy in Huizhou after reading Su's verse[3]
1100	63	Returned to mainland, with permission to live anywhere
1101	64	Died of illness in Changzhou, Aug. 24, with entire family by his side

[1][Yang, 2006, p. 59]. [2][Yang, 2006, p. 83]. [3][Lin, 1947, p. 368].

Su was a very smart person. He was not a dogmatist in philosophies, unlike many of his fellow countrymen. He exercised judgment in selecting and applying whatever he found suitable from Confucianism, Taoism and Buddhism. Confucianism helped him in handling interpersonal relationships. Taoism kept him calm, in lieu of psychotherapists, when he was banished from the capital (to nice places like Hangzhou and not so nice places like Huizhou); and directed him to embrace nature. Buddhism smoothed out his agonies when his three wives died and when unpleasant things were encountered.

And Su was a very good scholar, innovative and flexible when needed. The excellence of his poems and essays had been well recognized since his early age. After he was released from jail at age 42, he stopped writing poems that got him into trouble.[3] Instead, he turned to writing Song prose. Breaking with tradition, he expanded the scope of prose to cover anything and everything that he felt for—nature, daily life, festivals, expired wives and other women, philosophical thoughts, and so on. All were very touchingly and skillfully done. Su became a pioneer and the foremost person as Song prose is concerned.

Furthermore, throughout his life, Su was surrounded by very gifted friends. Su, together with the younger Mi Fei, established the impressionistic style of Chinese painting, called *Scholar Painting* (initiated by Wang Wei in Tang Dynasty) [Jiang, 1993, pp. 162-163]. And bamboo is one of the major themes in their paintings.

17.2.2 *Su's Art View*

Here, we examine his thoughts and attitude on art and literature around the time of his exile. As he opposed the political reform by Wang An-Shi, he was impeached for writing many verses which were regarded as lese-majesty. He was exiled to Huangzhou in 1080. Though he did not write many poems at this time, he devoted himself to painting. Through this experience he began to realize that both painting and composing poems were originated from the same source of creation. Su's art view can be sum up by four major points.

1. <u>Both Tao and art are integral for painting</u>

All paintings are unconscious reflection of a philosophy. Su was heavily influenced by his mother's love for Tao. One of Zhuang Zi's well-known quotations, "Heaven and Earth coexist with me and everything is combined with me," can be considered as the basis of his view on painting bamboos. Su was totally aware that Tao itself cannot bring art to perfection. He realized there are men who can possess both Tao and art;

---

[3] After all, poems reached its zenith in the Tang Dynasty (618-907) or even before. And the very form of poems allowed his enemies to read his mind too easily.

others who possess Tao but not art—although the subject takes shape in their heart, it does not take shape in their hands that hold the brushes [Sirén, 1935, p. 440]. In fact, it is never easy to make the hand to cooperate with the heart. The spirit of the subject can only be fully expressed when the necessary technical skills have been mastered thoroughly by the painter.

2. <u>Painting is to bring out the "inherent reason" behind the subject</u>

Su objected to painting bamboo with too many leaves, because that would be just a copy of the object. At that time and even now, lots of people judged a painting only by its resemblance to the real object. Su did not agree with this criterion. Su believed the purpose of a painting is to bring out the "*inherent reason*" (*li*) behind the subject [Sirén, 1935, pp. 439-441]. Su is regarded as the first one who extracted the noumenon from the conventional conception theory [Leng, 2004, p. 500].

3. <u>Rapid rhythmic strokes done with a unifying conception</u>

The style of "scholar painting" emphasizes rapid rhythmic strokes done with a unifying conception of the subject [Lin, 1947, p. 274]. To Su and other scholar-painters, painting served mainly as a symbolic means of expressing visual ideas or reflexes of the mind [Sirén, 1935, p. 434].

Lin Yu-Tang believes art are all about rhythm, whether in painting, sculpture, or music. As long as beauty is movement, every art form has an implied rhythm [1947, p. 279]. In this sense, the brush movement follows Su's mind and the nature's rhythms can easily be seen from the brush movement. The lines and contours are results of a process of growth and serve a definite purpose [1947, p. 280]. The pictures that contain the nature's rhythms may be considered the impressionistic school of Chinese art, in which the artist is much more concerned about expressing in a controlling rhythm than making copy of the objects. Su described the painting process as "Then grasp the brush, fix your attention, so that you see clearly what you wish to paint; start quickly, move the brush, follow straight what you see before you, as the buzzard swoops when the hare jumps out. If you hesitate one moment, it is gone" [Su, 1986, p. 365].

Su sometimes painted after drinking a lot of wine. To him, wine kindles inspiration and improvement of brush speed; he is then no longer bound by any imposed rules. Wine is just the method that leads Su to the purified world, from which he can get spiritual energy. However, it cannot guarantee him to reach the perfect state for painting. Only a small number of bamboo paintings created by Su while he was a little drunk were preserved.

4. Absence of irrelevant matter

Su's painting of a few sprays of bamboo leaves with a barely visible moon shining from behind creates two effects. First, the absence of irrelevant matter stimulates the imagination of the spectator; second, it implies that these few bamboo leaves are worth looking at forever and ever in the delight of the simple rhythms they express [Lin, 1947, p. 281].

17.2.3 *Su's Bamboo Painting*

Su's bamboo painting is done with three guiding principles.

1. Bamboo is the emblem of hermits

Bamboo, considered as one of the noble plants by the Chinese, expresses a peculiar combination of "suppleness and strength" [Sirén, 1935, p. 437]. The thin bamboos are like hermits [1935, p. 441]. Their power to remain green even in the cold season and their habit of yielding and bending before the storm, without breaking and always coming back, appeal to Su and Chinese intellectuals like him. Yielding and bending can suggest life, and life can easily be seen from Su's unique bamboos.

2. Bamboo has no constant form but inner spirit

Some of Su's bamboos were painted in one stroke from the ground up to the top without any bamboo joints. For Su, a painting's ultimate value is justified by how well it satisfies the painter's mind, not by the painting skills or the highly-finished painting itself, although he once had a great admiration for this type of work. He fully expressed this view in two lines: "To judge a painting by its verisimilitude is to judge it at the mental level of a child" [Su, 1986, p. 367]. Fewer details make it easier to express nature's rhythm. Correct proportions and vividness are not as important as the "*constant principle*" [1986, p. 367], which was interpreted as the "inner spirit" by Lin Yu-Tang [1947, p. 282]. The constant principle, or in other words, inner spirit, is the rule of nature.

3. Constant form and constant principle are against each other

The bamboos that Su painted have no *constant form* but inner spirit. "The lack of constant form does not spoil the whole thing, but the whole thing would lose spiritual significance without the inner spirit." [Su, 1986, p. 367]. Without the constant form, one can break away from the material form and rule, so that he can truly grasp the reason for the nature. For Su, the lack of constant form is "the constant form," which embodies the rule of nature. Su kind of regarded the constant form and the constant principle as opposing each other, with the result that some of his bamboos (Fig. 17.2) are thinner than usual, the bamboo leaves are not always linked to the stem, and the bamboo joints' bulging parts are less tangible, unlike the bamboos painted by his contemporaries such as Wen Tong (Fig. 17.3).

17.3 Paul Cézanne and His Apple

Paul Cézanne's career, art view and painting innovations are presented here.

17.3.1 *Cézanne's Life and Career*

Paul Cézanne[4] was born on January 19, 1839 in Aix-en-Provence in southern France. His father is a well-to-do banker who supported Cézanne financially all his life; unlike other impressionist painters of his time, Cézanne had no need to worry about selling his paintings.

Upon the insistence of his father, Cézanne studied law for three years while taking classes at the drawing academy. At age 22, he went to Paris to study painting, with his father's permission, but was unhappy

[4] A short and very readable account of Paul Cézanne's life and art is presented in [Jennings, 1986, pp. 41-52], with color pictures; a long account is [Lindsay, 1969]. A useful reference is: http://en.wikipedia.org/wiki/Paul_C%C3%A9zanne. The one-hour long video *Cézanne in Provence* (pbs.org) is enjoyable.

there. He returned home shortly and worked in his father's bank; could not stand it and went back to Paris in 1862. He went home again, for good, in 1870; the motivation was to evade the military draft [Jennings, 1986, p. 44]. Cézanne kept his marriage and a son secret to his father for ten years, but not to his mother, for fear of offending the father.

Fig. 17.2. Su Dong-Po, *Bamboos and Stones*, 106 x 28 cm (partially shown).

Fig. 17.3. Wen Tong, *Ink Bamboo*, 131.6 x 105.4 cm.

While in Paris, Cézanne was in the company of the other impressionists but insisted to paint differently. It was Camille Pissarro (1830-1903), a father figure to Cézanne, who introduced him to out door painting; and Cézanne loved it for the rest of his life.

His joined the first Impressionist exhibition in 1874, skipped the second, and rejoined the third. He kept sending his paintings to the Salon, *the* annual government-sponsored juried exhibition in Paris [Romano, 1996, p. 27], and was rejected each time, until the year 1882, at age 43.

It was in his fourties that Cézanne found his own style of painting; he flourished and received wide recognition in his fifties. He loved his art so much that he wrote, "I have sworn to myself to die painting..." [Cézanne, 1995, p. 330]. And he got it. On Monday, October 15, 1906, Cézanne, age 67, was out painting as usual. But he got soaked in the rain for several hours, and was brought home in a laundry cart by others. The *next* day, he went to the garden to work on a portrait, and then went home collapsed. Six days later, Cézanne died at home [Lindsay, 1969, pp. 341-342].

Cézanne, together with Seurat, Gauguin, van Gogh and others, pioneered post-impressionism [Strickland, 2007, p. 112], but it was Cézanne who opened the door to modern painting; in fact, modern art (see Section 1.3.2). Pablo Picasso (1881-1973) praised him as "my one and only master." Both Picasso and Henri Matisse (1869-1954) called him "the father of us all."[5]

17.3.2 *Cézanne's Art View*

Cézanne hardly wrote anything on his art view. But in the letters (1889-1906) to his young friends, especially the painter Emile Bernard (1868-1941), he did express some of his thoughts on this subject [Cézanne, 1995]. Here are some quotes from these letters and elsewhere.[6]

<u>On the nature of painting</u>

- Painting from nature is not copying the object; it is realizing one's sensations.
- Pure drawing is an abstraction. Drawing and color are not distinct; everything in nature is colored.
- The man of letters expresses himself in abstractions whereas a painter, by means of drawing and color, gives concrete form to his sensations and perceptions. (p. 303)

<u>On painting nature</u>

- The truth is in nature, and I shall prove it.
- In order to make progress, there is only nature, and the eye is trained through contact with her. (p. 306)

---

[5] http://en.wikipedia.org/wiki/Paul_C%C3%A9zanne (May 12, 2010).

[6] http://mypaulcezannepainting.com/Blog/category/quotations-from-paul-cezanne (May 12, 2010).

On painting method

- There are two things in the painter, the eye and the mind; each of them should aid the other.
- A work of art which did not begin in emotion is not art.
- Let us go forth to study beautiful nature, let us try to free our minds from them, let us strive to express ourselves according to our personal temperament. (p. 315)
- I believe in the logical development of everything we see and feel through the study of nature and turn my attention to technical questions later. (p.330)
- Treat nature by means of the cylinder, the sphere, the cone. (p. 301)

On painting assessment

- Taste is the best judge. It is rare. Art addresses itself only to an excessively limited number of individuals. (p. 302)
- When I judge art, I take my painting and put it next to a God object like a tree or flower. If it clashes, it is not art.
- Michelangelo is a constructor, and Raphael an artist who, great as he may be, is always tied to the model.—When he tries to become a thinker he sinks below his great rival. (p. 309)

On being a painter

- Talking about art is almost useless. (p. 303)
- The Louvre is a good book to consult but it must be only an intermediary. The real and immense study to be undertaken is the manifold picture of nature. (pp. 302-303)
- Genius is the ability to renew one's emotions in daily experience.

From these quotes, one can see that Cézanne considered that

- Nature is the most worthy as painting subjects are concerned.
- The painter should start with his emotion and depict what he feels (not the impression) about the subject, copying the subject is out of the question.
- Those who can appreciate good art are rare.

- Some great artists may not be so great in his eye.[7]

17.3.3 *Cézanne's Apple Painting*

Cézanne once said, "With an apple I will astonish Paris."[8] Well, he did not really paint an apple significantly different from others before him. But he did find a new way to paint a set of apples, by arranging them (usually with other fruits and other objects on a table) in such a way that different subsets of the arrangement are shown in different perspective. In other words, *multiple perspectives* coexist in the same painting, breaking the single-perspective tradition.[9]

Fig. 17.4. Paul Cézanne, *The Kitchen Table* (1888-1890).

[7] Belittling fellow colleagues, past or present, is a common trait in all creative professions. For example, if you worship Einstein, you probably could not become a great physicist; the reason is pretty complicated (see [Tsui & Lam, 2010]).

[8] http://mypaulcezannepainting.com/Blog/category/quotations-from-paul-cezanne (May 12, 2010).

[9] *Mona Lisa* by Leonardo da Vinci (1452-1519) is a famous example of the single-perspective design, in which all lines converge to a point behind Mona Lisa's head [Strickland, 2007, p. 34].

Figure 17.4 is an example of this multi-perspective design. For instance, the basket handle is shown from the perspective of looking from the right side of the picture, while the jar and the pear on the right front is seen from the front [Loran, 1943]. This fusing of different viewpoints on a single canvas was used by Picasso [Miller, 1996, p. 414].

The second innovation by Cézanne is equally important. To render what he saw as the *essence* behind the subject matter, neither the subject's shape nor the painter's impression, he structurally ordered whatever he perceived into simple forms and color planes.[10] This is what he meant when he wrote, "Treat nature by means of the cylinder, the sphere, the cone." This technique was adopted and pushed to the limit by Picasso in his Cubism paintings [Miller, 1996].

In fact, near the end of his life, Cézanne's paintings got simpler and simpler and became more abstract, like the one shown in Fig. 17.5, which is at the brink of abstract painting.

Fig. 17.5. Paul Cézanne, *Mount Sainte-Victoire Seen from Lauves* (1904-1905).

[10] http://en.wikipedia.org/wiki/Paul_C%C3%A9zanne (May 12, 2010).

17.4 Neurological Basis of Their Artworks

That Su Dong-Po's detached bamboo leaves and Paul Cézanne's color-plane brushes, as well as his multi-perspective composition, work is due to the complex mechanism of seeing in the viewer's brain. The old understanding of vision is that our brain simply tells us how the pixels of light received by the eye are spatially arranged. The new understanding based on neurological studies tells a different story [Kolak et al, 2006, pp. 82-86; Livingstone, 2002].

Essentially, light enters the cornea, focused by the lens and projected to the retina, where photoreceptors convert light signals into electric signals. The signals are processed first in the occipital lobe (at the back of the brain) and then the parietal lobe and the temporal lobe (see Fig. 17.6). In so doing, the brain analyses the signals, recalls information from memory for comparison, presents a final interpretation of the image that the eye receives, and excites other neurons as a response.

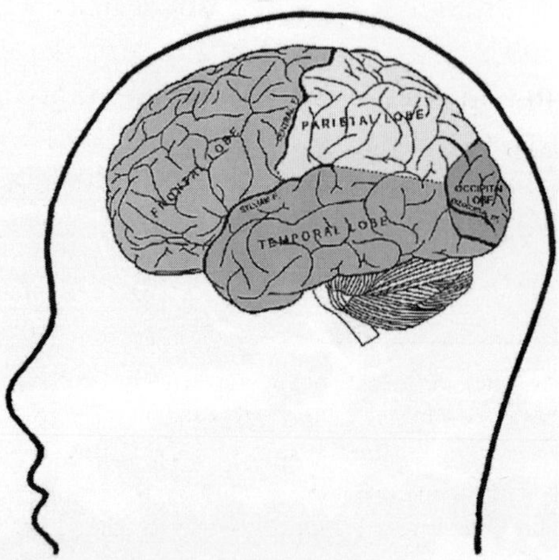

Fig. 17.6. The brain lobes: Frontal (left), parietal (middle up), temporal (middle down) and occipital (right).

Since we do know from observing nature that all tree leaves are connected to the stems, our brain was able to interpret Su's detached leaves correctly as part of the bamboo and not something that flow in air.

The same goes for Cézanne's paintings. That is, our brain connects the dots for us. And since the brain of each person is different, it follows that no two persons see the same thing identically; minuscule difference in their interpretation and appreciation of the same artwork is always present.

17.5 Discussion and Conclusion

It is interesting to compare and contrast the life and career paths as well as the subsequent influence of these two giants in the humanities, Su Dong-Po from the East and Paul Cézanne from the West.

1. Su made his living as a public servant[11] while Cézanne was supported mainly by his father. This shows that it is important to have financial security as an artist. Moreover, Su's stormy and dramatic career allowed him to move around the country with plenty of spare time, which benefited greatly his arts.

2. That two painters living almost a thousand years apart from each other with very different cultural backgrounds came to the same conclusion that the aim of painting is to depict the essence behind the object and not the appearance or impression of it illustrates the same human nature that the two shared, in consistent with the slow variation of human nature according to Darwin's evolution theory and the out-of-Africa understanding that all human beings shared the same ancestors who came out of Africa about 60,000 years ago (see [Lam, 2010]).

---

[11] Su was not a rebel and definitely not a maverick in politics. He went through the national examinations and worked in the government for two reasons: First, being an officer was the only way for an intellectual to get a decent living in a feudal society. Second, according to Confucius teaching, the foremost duty of an intellectual is to serve his country. To that end, the most efficient way was getting oneself near the emperor and let him listen to all your smart advices; less than that, became the governor of someplace oneself and served the people directly. Unfortunately, the ever changing political climate of the Song Dynasty was beyond what Su's limited political skills could handle. The good news for Su is that no other political figures survived the roller-coaster Song court much better than he did. The reason is that Northern Song emperors came and gone frequently (9 emperors in 168 years), like the rapid shuffling of CEOs in a big, modern company—and the stock price plunges, with one important difference; i.e., arts flourished in the former as a result but not in the latter.

3. As for the bamboo painting style that Su initiated, unfortunately, with few exceptions, it was not adopted by other Chinese painters. This is in sharp contrast to the fate of Cézanne painting style, which opened up the era of modern arts in the West. The reason behind this difference in fate is due to the driving force of science and technology that existed in Cézanne's Europe but not in Su' China. In Europe, color photography appeared in 1873, thirty six years after photography was invented by Daguerre [Strickland, 2007]. These science/technology advances made the birth of Impressionism almost inevitable. (In fact, the first official Impressionism exhibition was held in 1874.) On the other hand, China was an ultra-stable society [Jin & Liu, 1992; 1993], in which innovation was discouraged by Confucianism. Furthermore, mathematics was not included in the national examination in old China in spite of the few ingenious mathematical achievements. This resulted in the lack of interest and advance in science in the society. Arts in China suffered, unfortunately for Su's bamboo.

Appendix 17.1: Dong-Po Pork

There is no absolute proof that Dong-Po Pork (Fig. 17.7) was indeed invented by Su.[12] However, the following poem, embodied in *Complete Works of Su Shi* [Su, 2000, p. 1047], could be considered as a sort of "proof."

[12] The folklore has two versions: (1) One day, Su was cooking stewed pork himself (cooking is not an unusual habit for some Chinese male artists, not to mention the Italians), a good friend came in for a visit. The two played Chinese chess for several hours, and Su forgot his cooking. When he remembered it, the pork had been over cooked, but it actually tasted better, and Dong-Po Pork was born. [Source: http://en.wikipedia.org/wiki/ Su_Shi (March 3, 2010).] (2) Su directed the building of Su Causeway in 1090, Hangzhou, which favored the citizens. Su received so many pork in return that he cooked Dong-Po Pork and distributed it to the workers; see below for more details. (Source: Display in Su Dong-Po Memorial Hall, Hangzhou.) These two tales could both be right but are never verified.

Fig. 17.7. A piece of Dong-Po Pork. [Photograph taken by Lui Lam, at Grandma's Kitchen, Beijing, May 24, 2009.]

Ode to Pork

Ensure continued low burning with a little water after washing the pot clean.
The flavor of the pork is amazing if the lid is lifted at the right time.
The price of pork in Huangzhou is dirt cheap.
The rich disdain eating it, and the poor don't really know how amazing it could be with the right cooking methods.
I eat two bowls of pork after getting up every morning.
Leave me alone; I am stuffed.

It is obvious that *Ode to Pork* was written in Huangzhou, not Hangzhou. Su stayed in Huangzhou for four years from 1080 and went again to his post in Hangzhou in 1089. The recipe could be brought to Hangzhou by Su and ended up becoming one of the best-known Hangzhou dishes.

Both *Records of the West Lake* [Shi, 1995] and *The Integration of West Lake Documents* [Wang, 2004] embody the specialty products of Hangzhou. However, Dong-Po Pork is recorded in neither of these two books. He Yin-Jie, the guide in Su Dong-Po Memorial Hall, stated positively that no existing official history book on Hangzhou records

Dong-Po Pork.[13] As a matter of fact, Dong-Po Pork can be found at many places where Su ever stayed, such as the provinces of Sichuan, Jiangxi, Jiangsu, and Hubei. These dishes share the same name, but the recipes are different. The foregoing poem named *Ode to Pork* is the only one of all Su's works that explain definitely how to cook pork. This clue leads to the days that Su was in Huangzhou, Hubei Province. Rao Xue-Gang, who studied Su's life in Huangzhou for over 23 years and published his research results as *Su Dong-Po in Huangzhou* in 1999, confirmed[14] that *Ode to Pork* was written in Huangzhou in April, 1082.[15]

Although *Ode to Pork* was written in Huangzhou, the pork was not named Dong-Po Pork yet. Su loved "stewed pork with brown sauce,"[16] but Dong-Po Pork was not named by Su himself. In *Ode to Dong-Po Thick Soup*, Su brought forth "Dong-Po" definitely in the title and the poem: "Dong-Po thick soup, is the vegetable thick soup cooked by Dong-Po Hermit" [Su, 2000, p. 1046].

In the folklore of Dong-Po Pork's Hangzhou version, Su received a lot of rice wine and pork from graceful local people after he directed the building of Su Causeway. He told his family to cook "stewed pork with

---

[13] Interview of He Yin-Jie by Li-Meng Qiu, Su Dong-Po Memorial Hall, May 18, 2010.

[14] Telephone interview of Rao Xue-Gang by Li-Meng Qiu, May 26, 2010.

[15] The name of the poem listed in the two chronologies of Su in Rao's book is *Ode to Boiling Pig's Head* [Rao, 1999, p. 17 & p. 157]. Whereas the name of the same poem embodied in one of the articles of the book is *Ode to Pork* [Rao, 1999, p. 331]. Rao confirmed *Ode to Pork* was called *Ode to Boiling Pig's Head* as well. If it is true, the main ingredient of cooking Huangzhou Dong-Po Pork was quite different from that of Hangzhou Dong-Po Pork while the poem was written, because the latter uses streaky pork. *Su Shi Corpus* [Su, 1986] and *Poetry of Su Shi* [Su, 1982] embody this poem as *Ode to Pork*. According to the footnote to *Ode to Pork* in *Su Shi Corpus*, this poem has another name, *Ode to Boiling Pork Thick Soup* [Su, 1986, p. 597]. The same poem is embodied as *Eating Pork* in *Bamboo Slope Random Notes on Classical Poets and Poetry*, written by Zhou Zi-Zhi [He, 1981, p. 351]. Zhou was born in 1082 and also lived in Song Dynasty. *Eating Pork* is a little bit different in order and characters, but the meaning is the same. The only thing notable besides the name is, "I eat one bowl of pork after getting up every morning", not "two bowls of pork." Although this version might be more reliable, it is not the version widely adopted, probably because the approximate number used in Chinese language is usually expressed by "two".

[16] "Fu Yin cooks pork to treat Dong-Po" appears in *Answering Fu Yin Playfully*, written by Su in Huangzhou [1982, p. 2654]; it clearly shows that how much Su loved stewed pork.

brown sauce" and distribute pork to the workers who took part in the building. His family heard "cook pork with rice wine" by mistake. The local people were deeply touched by his generous action and named the pork after Dong-Po [Lin & Shen, 1993, p. 49]. This folklore can be considered as a circumstantial evidence that Su himself did not name "Dong-Po Pork." And it is quite notable that this recipe is different from the Huangzhou recipe, because "rice wine" takes the place of "water."

The cooking time is not explicit in the folklore. *Ode to Pork* mentioned "the flavor of the pork is amazing if the lid is lifted at the right time." "The right time" is present in *Scattered Records of the Fairy in Cloud*, edited by Feng Zhi. A man named Huangsheng cooked 1 kg venison from morning everyday. He would say it is the right time by nightfall [Feng, 2008, p. 75]. The original story was recorded in *Records of Cheng'an*, which is no longer extant. Feng lived in Later Tang Dynasty (923-936); Su might have read Feng's book.

The two best restaurants in Hangzhou (Lou Wai Lou and Zhi Wei Guan) are unwilling to reveal the recipe of Dong-Po Pork. The following is the recipe of Dong-Po Pork displayed at the Su Dong-Po Memorial Hall (May 18, 2010).

Ingredients

Streaky pork[17] 1500 g
Green onion 50 g
Knotted green onion 50 g
Refined white sugar 100 g
Shaoxing rice wine 250 g
Ginger 50 g
Soy sauce 150 g

The cooking

Dice the pork after washing.[18] Boil pork in boiling water for five minutes. Put the pork skin downwards into a big earthen pot with green

[17] The best ingredient for cooking Dong-Po Pork is the Jinhua pig (alias Liangtouwu pig), which has white trunk and black head and black rump. The pigskin is thin, so are the bones. Jinhua pig has been farmed since the West Jin Dynasty (266-316).

onion and ginger on bamboo grid. Get the lid on the pot after adding refined white sugar, soy sauce, Shaoxing rice wine and knotted green onion in turn. Tuck "peach blossom paper"[19] into the aperture between the lid and the earthen pot. Put the earthen pot over "high" heat. Adjust stove heat to "low" after the cooking liquid starts boiling. Turn off stove when pork is tender.[20] Skim fat and put the small clay pot with pork skin upwards in a bamboo steamer. Adjust stove heat to "high" and steam for 30 minutes.

As set forth, the recipe of Dong-Po Pork is different at different places. The way of cooking Hangzhou Dong-Po Pork is a typical localism method. Yuan Mei (1716 – 1797), a well-known poet, scholar and artist in Qing Dynasty, was born in Hangzhou. He is an epicure and an expert on Hangzhou Style of cooking. What is interesting is that unlike Su, Yuan himself did not cook. However, his book *Sui Yuan Cooking Methods* does describe the way to cook pork. According to Yuan, the cook should be patient and cook pork with the skin downwards first. Oil and fat will soak through the skin. It is the way to keep the flavor and make the skin tastes crisp. Otherwise the skin will get hard and the flavor will disappear [Yuan, 2000, p. 26]. Yet, Yuan did not mention Dong-Po Pork in his book. All he recorded are the "universal" methods of cooking.

References

Cézanne, P. [1995] *Paul Cézanne Letters*, ed. Rewald, J. (Da Capo, New York).

Feng, Zhi (ed.) [2008] *Scattered Records of the Fairy in Cloud* (Zhonghua Book Company, Beijing).

Jennings, G. [1996] *Impressionist Painters* (Chancellor Press, London).

Jiang, Xun [1993] *History of Chinese Art* (East China Books, Taibei).

---

[18] According to the recipe in *West Lake Cyclopedia* [Shen & Zhang, 2005, p. 177], the diced pork should be 75 g each.

[19] Peach Blossom Paper, thin and tough, is a type of semitransparent paper used for making window paper and kites.

[20] According to the recipe in *West Lake Cyclopedia* [Shen & Zhang, 2005, p. 177], it takes 2 hours to get the pork tender on simmer.

Jin, Guan-Tao & Liu, Qing-Feng [1992] *The Cycle of Growth and Decline: On the Ultrastable Structure of Chinese Society* (Chinese University of Hong Kong, Hong Kong).

Jin, Guan-Tao & Liu, Qing-Feng [1993] *The Transformation of Chinese Society (1840-1956): The Fate of Its Ultrastable Structure in Modern Times* (Chinese University of Hong Kong, Hong Kong).

Kolak, D., Hirstein, W., Mandik, P. & Waskan, J. [2006] *Cognitive Science: An Introduction to Mind and Brain* (Routledge, New York).

Lam, L. [2010] "Arts: A Science Matter," in *Arts: A Science Matter*, eds. Burguete, M. & Lam, L. (World Scientific, Singpapore).

Leng, Cheng-Jin [2004] *Su Shi's View on Philosophy and Art* (Xueyuan Press, Beijing)

Li, Yi-Bing [1996] *A New Biography of Su Dong-Po* (Linking, Taibei).

Lin, Yu-Tang [1947] *The Gay Genius: The Life and Times of Su Tungbo* (John Day, New York).

Lin, Zheng-Qiu & Shen, Guan-Zhong (eds.) [1993] *Hangzhou, China, Lou Wai Lou* (Zhejiang Photography Press, Hangzhou).

Lindsay, J. [1969] *Cézanne: His Life and Art* (New York Graphic Society, New York).

Livingstone, M. [2002] *Vision and Art: The Biology of Seeing* (Abrams, New York).

Loran, E. [1943] *Cézanne Composition* (Univ. of California Press, Berkeley).

Miller, A. I. [1996] *Insights of Genius: Imagery and Creativity in Science and Art* (Copernicus, New York).

Rao, Xue-Gang [1999] *Su Dong-Po in Huangzhou* (Jinghua Press, Beijing).

Romano, E. [1996] *The Impressionists* (Penguin Studio, New York).

Shen, Guan-Zhong & Zhang, Wei-Lin (eds.) [2005] *West Lake Cyclopedia: Lou Wai Lou Restaurant* (Hangzhou Press, Hangzhou).

Shi, Ji-Dong (ed.) [1995] *Records of the West Lake* (Shanghai Ancient Book Press, Shanghai).

Sirén, O. [1935] "Su Tung-p'o as an art critic," *Geografiska Annaler*, Vol. 17, Supplement: Hyllningsskrift Tillagnad Sven Hedin, pp. 434-445.

Strickland, C. [2007] *The Annotated Mona Lisa: A Crash Course in Art History from Prehistoric to Post-Modern* (Andrews McMeel, Kansas City).

Su, Shi [1982] *Poetry of Su Shi*, ed. Wang Wen-Gao (Zhonghua Book Company, Beijing).

Su, Shi [1986] *Su Shi Corpus* (Zhonghua Book Company, Beijing).

Su, Shi [2000] "Ode to Pork," in *Complete Work of Su Shi*, Vol. **3**, punctuated by Fu Cheng & Mu Chou (Shanghai Ancient Book Press, Shanghai).

Tsui, H. & Lam, L. [2010] "Making movies and making physics," in *Arts: A Science Matter*, eds. Burguete, M. & Lam, L. (World Scientific, Singapore).

Wang, Guo-Ping (ed.) [2004] *The Integration of West Lake Documents* (Hangzhou Press, Hangzhou).

Watson, B. [1994] *Selected Poems of Su Tung-p'o* (Copper Canyon, Port Townsend, WA).
Yang, Pin [2006] *Su Dong-Po* (Shanxi Educational Press, Taiyuan).
Yuan, Mei [2000] "Boiling Pork," in *Sui Yuan Cooking Methods* (Jiangsu Ancient Book Press, Nanjing).
Zhou, Zi-Zhi, [1981] "Eating Pork," in *Bamboo Slope Random Notes on Classical Poets and Poetry, All Ages Random Notes on Classical Poets and Poetry*, ed. He Wen-Huan (Zhonghua Book Company, Beijing).

Acknowledgments

About two thirds of the chapters in this book are expanded, written versions of selected invited talks presented at the Second International Conference on Science Matters, *Arts & Science*, held in Estoril, Portugal, October 5-7, 2009, and co-chaired by Maria Burguete and Lui Lam. The rest of the chapters are invited contributions specifically for this book after the conference.

The Local Scientific Committee of this conference consists of João Calvão, Roberto Carneiro, Bernardo Herold, Luís Portela, Claudina Rodrigues-Pousada and Fernando Ramôa Ribeiro. The Advisors are Paul Caro, Brigitte Hoppe, Maurizio Salvi, Michael Shermer and Edward O. Wilson. The sponsors are Fundação Luso-Americana, Fundação para a Ciência e Tecnologia, Fundação Calouste Gulbenkian, Câmara Municipal de Cascais, Casino Estoril, Centro dos Congressos Estoril and Instituto Rocha Cabral. And the conference was under the auspices of the International Science Matters Committee, members of which are: Maria Burguete (Portugal), Paul Caro (France), Patrick Hogan (USA), Brigitte Hoppe (Germany), Lui Lam (USA), Bing Liu (China), Dun Liu (China), John Onians (UK), Nigel Sanitt (UK), Ivo Schneider (Germany) and Michael Shermer (USA).

We are much grateful to all the individuals and organizations mentioned above that guaranteed the success of this second SciMat conference. We thank our personal friend, Lucinda Moraes, for assistance before and during the conference. Apart from these efforts, this book would not be possible without the cooperation and patience of the contributors. Furthermore, we heartily thank K. K. Phua, the Chairman and Chief Editor of World Scientific, for his wise decision in establishing this new book series, Science Matters Series, at the early development stage of this new discipline, which takes real insight and courage; and Kim Tan, the editor from WS overseeing the publication of the books in this series, for her unfailing support and superb efficiency.

FUNDAÇÃO CALOUSTE GULBENKIAN

INSTITUTO DE INVESTIGAÇÃO CIENTÍFICA BENTO DA ROCHA CABRAL

FCT Fundação para a Ciência e a Tecnologia
MINISTÉRIO DA CIÊNCIA, TECNOLOGIA E ENSINO SUPERIOR

Cascais Câmara Municipal

CASINO ESTORIL

estoril MEETINGS

FUNDAÇÃO LUSO-AMERICANA

Contributors

Maria Burguete received her Ph.D. in History of Science (contemporary chemistry) from Ludwig Maximilians University at Munich, Germany (2000). She was the very first biochemist to graduate from the Faculty of Sciences in Lisbon (1982), after completing a Bachelor's Degree in Chemical Engineering (1979) at the Lisbon Higher Institute of Engineering (ISEL). She is a scientist and a university lecturer with teaching and research experience in a wide variety of scientific fields. Burguete is now a scientific researcher at Scientific Research Institute Bento da Rocha Cabral in Lisbon, Portugal, and a Correspondent Member of European Academy of Sciences, Arts and Letters. She has published five books in the scientific field (and eight books of poetry) and over 30 scientific papers. *Email: mariaburguete@gmail.com.*

D. Fox Harrell is a researcher exploring the relationship between imaginative cognition and computation. He is an Assistant Professor of Digital Media at the Georgia Institute of Technology. He directs the Imagination, Computation, and Expression [ICE] Lab/Studio (icelab.lcc.gatech.edu) in developing new forms of computational narrative, gaming, social networking, and related technical-cultural media based in computer science, cognitive science, and digital media arts. The National Science Foundation has recognized Harrell with an NSF CAREER Award for his project "Computing for Advanced Identity Representation." Harrell holds a Ph.D. in Computer Science and Cognitive Science from the University of California, San Diego. He also earned a B.F.A. in Art, a B.S. in Logic and Computation, and minor in Computer Science at Carnegie Mellon University, each with highest honors. He has worked as an interactive television producer and as a game designer. *Email: fox.harrell@lcc.gatech.edu.*

Lalita Pandit Hogan received her M.A. and Ph.D. in English from the State University of New York at Buffalo, New York. She is a professor of English at the University of Wisconsin-La Crosse, where she teaches Shakespeare, Critical Theory and World Literature. She has published articles and book chapters on Shakespeare, Tagore, Goethe, Comparative Aesthetics and Indian Cinema. She is co-editor of three books and three special issues, which include *Rabindranath Tagore: Universality and Tradition* (2003); *Cognitive Shakespeare*: *Criticism and Theory in the Age of Neuroscience* (Winter 2006) and *Hindi Cinema,* special issue of *Projections: Journal of Movies and the Mind* (Winter 2009). *Email: hogan.lali@uwlax.edu.*

Patrick Colm Hogan received his B.A. in Philosophy from the University of Santa Clara, his M.A. in Philosophy from the University of Chicago, and his Ph.D. in English from the State University of New York at Buffalo. He is a professor in the Department of English as well as the Program in Comparative Literature and Cultural Studies and the Program in Cognitive Science at the University of Connecticut. He is the author of 12 books—including *The Mind and Its Stories: Narrative Universals and Human Emotion* and *Cognitive Science, Literature, and the Arts: A Guide for Humanists*—and over 100 professional articles. He is currently completing work as editor of *The Cambridge Encyclopedia of the Language Sciences. Email: Patrick.Hogan@uconn.edu.*

Brigitte Hoppe obtained her state diploma in pharmaceutical sciences from the University of Freiburg (Breisgau, Germany); she earned a Ph.D. in the History of Science from the University of Frankfurt (Main) in 1964. Her research on epistemological changes in life sciences during Early Modern Times was the basis of the post-doctoral thesis (habilitation) at the University of Munich in 1972, where she became Associate Professor in 1980. She pursued research (partially sponsored by research foundations) in national and international archives and is a member of national and international learned societies. Hoppe has published 7 books and over 200 papers in the field of the history of the sciences. Her current research deals the history of life sciences from the 16th to the 20th centuries. *Email: B.Hoppe@lrz.uni-muenchen.de*

Lui Lam obtained his B.Sc. (with First Class Honors) from the University of Hong Kong, M.Sc. from University of British Columbia, and Ph.D. from Columbia University. He is Professor of Physics at San Jose State University, California and Adjunct Professor at both the Chinese Academy of Sciences and the China Association for Science and Technology. Lam invented bowlics (1982), one of three existing types of liquid crystals in the world; active walks (1992), a new paradigm in complex systems; and a new discipline called histophysics (2002). He published 12 books and over 170 scientific papers. He is the founder of the International Liquid Crystal Society (1990); cofounder of the Chinese Liquid Crystal Society (1980); founder and editor of two book series, *Science Matters* and *Partially Ordered Systems*. His current research is in science matters, histophysics and complex systems. *Email: lui2002lam@yahoo.com.*

Holly Lane earned her B.F.A. in Art, with Great Distinction (1986) and her M.F.A. in Art—Pictorial Art (1988) from San Jose State University. Painter and sculptor, Lane has had 18 solo exhibitions at galleries and museums, and been included in 86 group exhibitions including an influential exhibition at the Whitney Museum of American Art at Champion. Her work is in 80 private collections and 10 public collections. Over 40 reviews and articles on her work have been published by magazines and newspapers: The *New York Times*, *The New Yorker*, *Art News and Art* in America, to name a few. The art history textbook by Terence Grieder, *Artist and Audience* (1996), highlights her work. Lane is currently preparing for her tenth solo show in New York City. *Email: hlane42@comcast.net.*

Cristina Leiria studied at the Graduate School of Fine Arts, Lisbon and Development Planning Unit, University College London. She was an architect in United Kingdom, Mozambique, Zimbabwe, South Africa, Portugal and Macao. From a young age she has spontaneously sculptured, and in 1992-1994, aside from architecture, started to develop this ability finding in the world of silence a new dimension of life. After 1999, trying to infuse urban life with hope and peace through the joining of spiritual, social and architectonic domains, her work focuses on Public Art such as the Kun Iam Ecumenical Center (1999), 22 m bronze statue at Lotus Flower Building, Macau, China; *Kun Iam and Lotus Flower* (2002), 2.7 m bronze, Lisbon; *Loving Birth* (2003), 3 m white-sea stone, Cascais; *Sailing the Wind* (2003), 12 m, Tavira; *Sailing* (2007), 3 m bronze, Cascais. *Email: crisrochlei@ gmail.com.*

Bing Liu obtained his B.Sc. from Peking University (physics department) and M.Sc. from Graduate School of Chinese Academy of Sciences. He is now a professor of history of science at Tsinghua University, vice director of the Center for Science Communication and Popularization of CAST and Tsinghua University, and Guest Professor at several universities in China including Shanghai Jiaotong University. He published 17 books (also translated 7 books and edited more than 30 books) and over 200 academic papers. Liu's research fields currently include history of physics, historiography of science, philosophy of science, and science communication. His Blog: http://blog.sina.com.cn/liubing1958. *Email: liubing@tsinghua.edu.cn.*

Linsen Hsia Ngai was born in China and grew up in the United States. She earned her B.S. in Chemistry from the University of California, Berkeley, and her M.S. and Ph.D. from University of Chicago majoring in Chemical Physics. She was a postdoctoral researcher at Northwestern University and Boston University before doing laser research experiments at Naval Research Laboratory. After retiring from scientific research she became a watercolorist. Her subject matters include animals, birds, wild lives, flowers and plants, landscapes, human figures and portraits. She is a member of the Mclean Art Society, Vienna Art Society, Potomac Valley Watercolorist, and the McLean Project for the Arts. Her arts are in display in many local shows and juried shows. Some of her arts were honored with awards and recognitions. Her artworks could be viewed at website: www.LStudios.org. *Email: sauhaar@yahoo. com.*

John Onians obtained his B.A. from Cambridge University and his Ph.D. from London University. From 1971 to 2007 he taught at the University of East Anglia, where he is now Emeritus Professor. He has also taught at several universities in the United States, at the Universities of Amsterdam and Leiden in the Netherlands and elsewhere. He has held fellowships at the Warburg Institute, London; the Centre for Advanced Research in the Visual Arts, Washington DC; the Getty Research Institute, Los Angeles; the Clark Art Institute, Williamstown and the Wissenschaftskolleg, Berlin. His books include *Art and Thought in the Hellenistic Age* (1979), *Bearers of Meaning* (1988) and *Neuroarthistory* (2007). He was founder editor of the journal *Art History* and edited the first *Atlas of World Art* (2004). He is now writing a neuroarthistory of Europe. *Email: j.onians@uea.ac.uk.*

Li-Meng Qiu obtained her B.A., M.A. and Ph.D. from Renmin University of China. She is assistant professor of Chinese Language at Zhejiang University, China. Qiu published 3 papers on simplified Chinese characters and traditional characters, prepositions of ancient Chinese language and excavated texts research. She taught Chinese Culture in the Confucius Institute at the University of Rhode Island (2009). Her current research is in teaching Chinese as a foreign language, ancient Chinese grammar and Chinese culture. *Email: qiulimeng202 @yahoo.com.cn.*

Nigel Sanitt obtained his B.Sc. in Physics from Imperial College, London and Part III of the Mathematics Tripos and Ph.D. from Cambridge University, where he trained as an astrophysicist at the Institute of Astronomy. He is founder and editor of *The Pantaneto Forum*, a journal which aims to promote debate on how scientists communicate, with particular emphasis on how such communication and research skills can be improved through a better philosophical understanding of science. His book *Science as a Questioning Process* was published in 1996, and he has edited a collection of articles from the first five years of *The Pantaneto Forum* under the title: *Motivating Science. Email: nigel@pantaneto.co.uk.*

Ivo Schneider is Professor emeritus for the history of science of the Universität der Bundeswehr München. He was professor and visiting professor at the universities of Munich, Princeton, Bielefeld, the university of Minnesota in Minneapolis and the technical university of Budapest, from which he got the degree of a Doctor honoris causa in 2004. Special research interests concern the history of classical probability theory, mathematical practitioners and reckoningmasters, scientific instruments in the 17th and 18th centuries, biographies of scientists, science theater and the origins of Bavarian optical industry. His books comprise a source book of the history of probability theory and biographies of Archimedes, Johannes Faulhaber and Isaac Newton. He is a member of different national and international societies for the history of science and ever since 1995 membre effectif of the Académie Internationale d'Histoire des Sciences. *Email: ivo.schneider@unibw.de.*

Hark Tsui was born in China, grew up in Vietnam and received secondary education in Hong Kong. He studied film at University of Texas, Austin, and earned a B.A. degree; then went to live and work in New York Chinatown from 1975-1977. Subsequently, he went back to Hong Kong and directed his first feature film *The Butterfly Murders* in 1979, ushering in the New Wave. Tsui has directed and produced over 70 movies, including *A Better Tomorrow* (1986), *A Chinese Ghost Story* (1987), *Swordsman* (1990), *Once Upon a Time in China* (1991), *A Chinese Ghost Story: The Tsui Hark Animation* (1997), *Seven Swords* (2005) and *All about Women* (2008). He was honored with, among others, the Golden Horse Award (1981), Hong Kong Film Award (1987, 1992) and the Outstanding Contribution to Asian Cinema Award (2009). Website: filmworkshop.net. *Email: fws@netvigator.com.*

Ting-Ting Wang obtained both her B.A. and M.A. in Chinese Literature from Renmin University of China. She has published research papers in journals, translated books and written novels. Her current interest is in film and mass culture studies. *Email: wtingting_fish@yahoo.com.cn.*

Guo-Sheng Wu obtained his B.Sc. and M.Sc. from Peking University, and Ph.D. from Graduate School of Chinese Academy of Social Sciences. He is Professor of History and Philosophy of Science, and director of Center for Social Study of Sciences at Peking University. He published more than 20 books and over 100 scientific papers. His current research is in phenomenological philosophy of science and technology, early modern history of mathematical physical sciences, and science communication. *Email: wugshpku@gmail.com.*

Jichen Zhu is an assistant professor of Digital Media in School of Visual Arts and Design at University of Central Florida, where she is the director of the Procedural Expression Lab. Her work focuses on developing humanistic and interpretive theoretical framework of computational technology, particularly artificial intelligence (AI), and constructing AI-based cultural artifacts. Her current research areas include digital humanities, software studies, computational narrative, and serious games. Zhu received a Ph.D. in Digital Media and a M.S. in Computer Science from Georgia Institution of Technology. She also holds a Master of Entertainment Technology from Carnegie Mellon University and a B.S. in Architecture from McGill University in Montreal, Canada. *Email: jzh@mail.ucf.edu.*

Index

C

D

E

G

H

M

N

S

T